Chemistry
A Guided Inquiry
Third Edition

Richard S. Moog
Franklin & Marshall College

John J. Farrell
Franklin & Marshall College

WILEY
JOHN WILEY & SONS, INC.

Cover Photo Credit: ©Royalty-Free/CORBIS

To order books or for customer service call 1-800-CALL-WILEY (225-5945).

ISBN 0-471-69941-1

Printed in the United States of America

10 9 8 7 6 5 4 3 2 1

Printed and bound by Malloy, Inc.

To the Instructor

These activities are designed to be used by students working in groups. There are many written materials available on-line to help instructors use these activities effectively. Please contact your Wiley representative for information on how to obtain access to these materials, or visit the web site at: http://www.wiley.com/college/moog.

The Process Oriented Guided Inquiry Learning (POGIL) project supports the dissemination and implementation of these types of materials for a variety of chemistry courses (high school, organic, physical, etc.). Information about the project and its activities (including workshops, additional materials, laboratory experiments) can be found at http://www.pogil.org. The POGIL project is supported by the National Science Foundation under Grant DUE-0231120.

Acknowledgments

This book is the result of the innumerable interactions that we have had with a large number of stimulating and thoughtful people.

- Special thanks to Dan Apple, Pacific Crest Software, for taking us to this previously untravelled path. The Pacific Crest Teaching Institute we attended provided us with the insights and inspiration to convert our classroom into a fully student-centered environment.

- Many thanks to Jim Spencer, Franklin & Marshall College, for his helpful and insightful discussions, comments, and corrections. We also thank Jim Spencer and his Spring 2004 Chemistry 112 students for using and evaluating a draft version of this third edition.

- We greatly appreciate the support and encouragement of the many members of the Middle Atlantic Discovery Chemistry Project, who have provided us with an opportunity to discuss our ideas with interested, stimulating, and dedicated colleagues.

- Thanks to the National Science Foundation (Grant DUE-0231120) for its support of the Process Oriented Guided Inquiry Learning (POGIL) project, which fosters the dissemination of guided-inquiry materials and encourages faculty to develop and use student-centered approaches in their classrooms.

- Thanks to the numerous colleagues who used our previous editions in their classrooms. Many provided us with insightful comments and suggestions for which we are grateful. We particularly note the contributions of Mary Hoppe and Joseph Byrne of Norwich University and Gary Edvenson of Minnesota State University, Moorhead.

- We thank Carol Strausser for her excellent work on this manuscript, constructing the figures and typing and re-typing our corrections and re-corrections.

- A great debt of thanks is due our students in General Chemistry at Franklin & Marshall College. Their enthusiasm for this approach, patience with our errors, and helpful and insightful comments have inspired us to continue to develop as instructors, and have helped us to improve these materials immeasurably.

Contents

To the Student

Science and engineering have dominated world events and world culture for at least 150 years. The blind and near blind have been made to see. The deaf and near deaf have been made to hear. The ill have been made well. The weak have been made strong. Radio, television and the internet have made the world seem smaller. And some of us have left the planet. Computers have played an essential role in all of these developments; they are now ubiquitous. These miraculous events happened by design—not by accident. Individuals and teams set out to accomplish goals. They systematically studied and analyzed the natural world around us. They designed and tested new tools. Human beings have embarked on a journey that cannot be reversed. We hope that you can participate in and contribute to these exciting times.

There is simply too much chemistry—not to mention physics, mathematics, biology, geology, and engineering—for any one person to assimilate. As a result, groups have become essential to identifying, defining, and solving problems in our society. This book was designed to be used by you as a working member of a group, actively engaged with the *important basic* concepts of chemistry. Our goals are to have you learn how to examine and process information, to ask good questions, to construct your own understanding, and to build your problem-solving skills.

If ever a book was written for students—this is it. This is *not* a textbook. This is *not* a study guide. This book is "a guided inquiry," in which you will examine data, written descriptions, and figures to develop chemical concepts. Each concept is explored in a *ChemActivity* comprising several sections—one or more **Model and Information Sections**, **Critical Thinking Questions,** and **Exercises and Problems**. You and your group study the Models and Information and systematically work through the Critical Thinking Questions. In doing so, you will discover important chemical principles and relationships. If you understand the answer to a question, but other members of your group do not, it is your responsibility to explain the answer. Explaining concepts to other members of your group not only helps in *their* understanding, it broadens *your* understanding. If you do not understand the answer to a question, you should ask one or more *good* questions (to the other members of your group). Learning to ask questions that clearly and concisely describe what you do not understand is an important skill. This book has many Critical Thinking Questions that serve as examples. To reinforce the ideas that are developed, and to practice applying them to new situations, numerous Exercises and Problems are provided; these are important for you to apply your new knowledge to new situations and solidify your understanding. We have found the combination of these methods to be a more effective learning strategy than the traditional lecture, and the vast majority of our students have agreed.

We hope that you will take ownership of your learning and that you will develop skills for lifelong learning. Nobody else can do it for you. We wish you well in this undertaking.

If you have any suggestions on how to improve this book, please write to us.

John J. Farrell
Richard S. Moog
Chemistry Department
Franklin & Marshall College
Lancaster, PA 17604
Rick.Moog@FandM.edu

ChemActivity 1

The Nuclear Atom

(What Is an Atom?)

Model: Schematic Diagrams for Various Atoms.

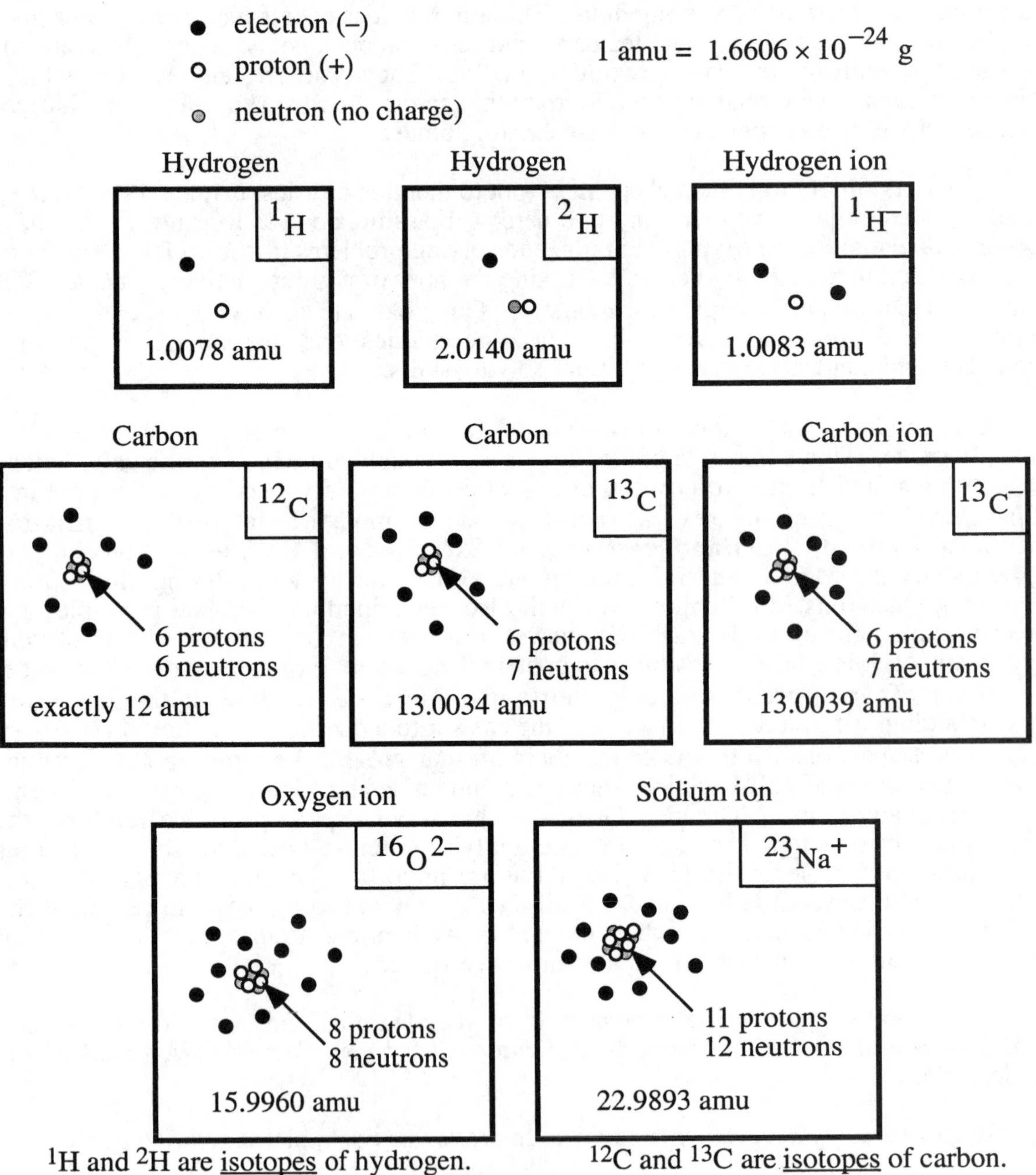

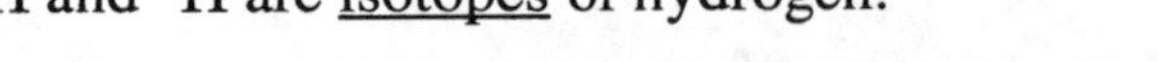

^{1}H and ^{2}H are isotopes of hydrogen. ^{12}C and ^{13}C are isotopes of carbon.

The nucleus of an atom contains the protons and the neutrons.

Critical Thinking Questions

1. How many protons are found in ^{12}C? ^{13}C? $^{13}C^{-}$?

6

2. How many neutrons are found in ^{12}C? ^{13}C? $^{13}C^{-}$?

6 7 7

3. How many electrons are found in ^{12}C? ^{13}C? $^{13}C^{-}$?

6 6 7

4. a) What feature distinguishes a neutral atom from an ion?

   b) How is the charge on an ion determined?

5. Based on the model,

   a) what do all carbon atoms (and ions) have in common?

   b) what do all hydrogen atoms (and ions) have in common?

   c) how many protons, neutrons, and electrons are there in one atom of $^{1}H^{+}$?

6. What is the significance of the atomic number, Z, above each atomic symbol in the periodic table?

7. Based on your answer to CTQ 6, what do all nickel (Ni) atoms have in common?

8. What structural feature is different in isotopes of a particular element?

9. How is the mass number, A, (left-hand superscript next to the atomic symbol as shown in the Model) determined (from the structure of the atom)?

10. Where is most of the mass of an atom, within the nucleus or outside of the nucleus? Explain your reasoning.

Exercises

1. Complete the following table.

Isotope	Atomic Number Z	Mass Number A	Number of Electrons
^{31}P	15		
^{18}O			8
	19	39	18
$^{58}Ni^{2+}$		58	

2. What is the mass (in grams) of a) one ^{1}H atom? b) one ^{12}C atom?

3. What is the mass (in grams) of 4.35×10^6 atoms of ^{12}C?

4. What is the mass (in grams) of 6.022×10^{23} atoms of ^{12}C?

5. What is the mass (in grams) of one molecule of carbon dioxide which has one ^{12}C atom and two ^{16}O atoms, $^{12}C^{16}O_2$?

6. a) Define mass number. b) Define atomic number.

7. Indicate whether the following statement is true or false and explain your reasoning.

   An ^{18}O atom contains the same number of protons, neutrons, and electrons.

8. How many electrons, protons, and neutrons are found in each of the following?

   ^{24}Mg $^{23}Na^{+}$ ^{35}Cl $^{35}Cl^{-}$ $^{56}Fe^{3+}$ ^{15}N $^{16}O^{2-}$ $^{27}Al^{3+}$

9. Complete the following table.

Isotope	Atomic Number Z	Mass Number A	Number of Electrons
	27	59	25
^{14}N			
	3	7	3
	3	6	3
$^{58}Zn^{2+}$			
$^{19}F^{-}$			

10. Using grammatically correct English sentences, describe what the isotopes of an element have in common and how they are different.

11. J. N. Spencer, G. M. Bodner, and L. H. Rickard, *Chemistry: Structure & Dynamics*, Third Edition, John Wiley & Sons, 2006. Chapter 1: Problems: 21, 24, 25, 29-31, 33, 49abd, 52.

Problems

1. Estimate the mass of one ^{14}C atom (in amu) as precisely as you can (from the data in the model). Explain your reasoning.

2. Use the data in Model 1 to estimate the values (in amu) of a) the mass of an electron, b) the mass of a proton, and c) the mass of a neutron.

3. The mass values calculated in Problem 2 are only approximate because when atoms (up through iron) are made (mainly in stars) from protons, neutrons, and electrons, energy is released. Einstein's equation $E = mc^2$ enables us to relate the energy released to the mass loss in the formation of atoms. Use the known values for the mass of a proton, 1.0073 amu, the mass of a neutron, 1.0087, and the mass of an electron, 5.486×10^{-4} amu, to show that the mass of a ^{12}C atom is less than the sum of the masses of the constituent particles.

ChemActivity 2

Atomic Number and Atomic Mass

(Are All of an Element's Atoms Identical?)

Model 1: Isotopes.

Each element found in nature occurs as a mixture of isotopes. The isotopic abundance can vary appreciably on an astronomical scale—in the Sun and on Earth, for example. On Earth, however, the abundance shows little variation from place to place.

Table 1. Natural abundance and atomic masses for various isotopes.

Isotope	Natural Abundance on Earth (%)	Atomic Mass (amu)
^{1}H	99.985	1.0078
^{2}H	0.015	2.0140
^{12}C	98.89	12.0000
^{13}C	1.11	13.0034
^{35}Cl	75.77	34.9689
^{37}Cl	24.23	36.9659
^{24}Mg	78.99	23.9850
^{25}Mg	10.00	24.9858
^{26}Mg	11.01	25.9826

1 amu = 1.6606×10^{-24} g

Critical Thinking Questions

1. How many isotopes of magnesium occur naturally on Earth?

2. Describe what all isotopes of magnesium have in common and also how are they different.

3. If you select one carbon atom at random, the mass of that atom is most likely to be __________ amu.

4. What is the mass (in amu) of 100 ^{12}C atoms? Of 100 ^{13}C atoms?

5. If you select one hundred carbon atoms at random, the total mass will be ____ .

   a) 1200.00 amu
   b) slightly more than 1200.00 amu
   c) slightly less than 1200.00 amu
   d) 1300.34 amu
   e) slightly less than 1300.34 amu

   Explain your reasoning.

   because carbon molecular weight is 12.01g per atom
   12.01 X 100 is slightly more than 1200 amu

Model 2: The Average Mass of a Marble.

In a collection of marbles, 25% of the marbles have a mass of 5.00 g and 75% of the marbles have a mass of 7.00 g. The average mass of a marble is 6.50 g.

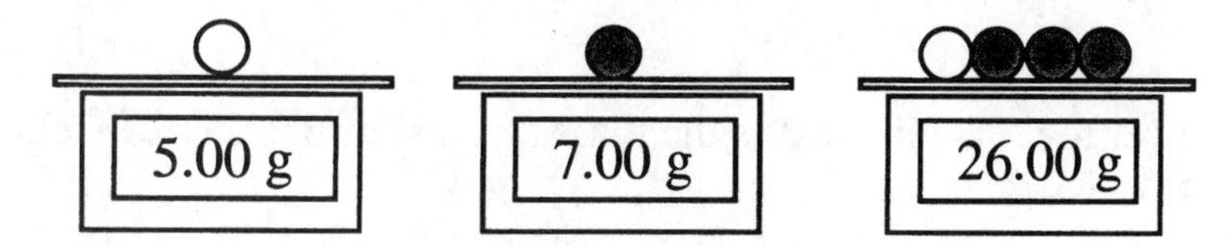

The average mass of a marble can be determined by dividing the total mass of the marbles by the total number of marbles:

$$\text{average mass of a marble} = \frac{1 \times 5.00\text{ g} + 3 \times 7.00\text{ g}}{4} = 6.50\text{ g} \qquad (1)$$

Or, the average mass of a marble can be determined by (a) multiplying the fraction of marbles of a particular type by the mass of a marble of that type and (b) taking a sum over all types of marbles:

$$\text{average mass of a marble} = 0.2500 \times 5.00\text{ g} + 0.7500 \times 7.00\text{ g} = 6.50\text{ g} \qquad (2)$$

Critical Thinking Questions

6. Show that equations 1 and 2 in Model 2 are equivalent by showing how the arithmetic expression in equation 1 can be transformed into the arithmetic expression in equation 2.

7. Do any of the marbles in Model 2 have the average mass?

8. a) Use the method of equation (2) in Model 2 to calculate the average mass of a chlorine atom in amu.

   b) Does any chlorine atom have this mass?

9. For any large collection of chlorine atoms (randomly selected):

   a) What is the average atomic mass of chlorine in amu?

   35.45

   b) What is the average mass of a chlorine atom in grams?

   35.45

10. Based on your answer to CTQ 9b, what is the mass (grams) of 6.022×10^{23} chlorine atoms (randomly selected)?

   35.45g

11. For a large collection of magnesium atoms (randomly selected):

   a) What is the average atomic mass of magnesium, Mg, in amu?

   b) What is the average mass of a Mg atom in grams?

12. Based on your answer to CTQ 11b, what is the mass (grams) of 6.022×10^{23} magnesium atoms (randomly selected)?

13. Examine the periodic table and find the symbol for magnesium.

    a) How does the number given just below the symbol for magnesium (rounded to 0.01) compare with the average mass (amu) of one magnesium atom?

    b) How does the number given just below the symbol for magnesium (rounded to 0.01) compare with the mass (grams) of 6.022×10^{23} magnesium atoms?

14. Find the symbol for chlorine on the periodic table.

    a) How does the number given just below the symbol for chlorine (rounded to 0.01) compare with the average mass (amu) of one chlorine atom?

    b) How does the number given just below the symbol for chlorine (rounded to 0.01) compare with the mass (grams) of 6.022×10^{23} chlorine atoms?

15. Give two interpretations of the number "12.011" found below the symbol for carbon on the periodic table.

16. Does <u>any</u> carbon atom have a mass of 12.011 amu? Does <u>any</u> magnesium atom have a mass of 24.305 amu?

Exercises

1. Without doing the calculations, what is the mass in grams of: a) 6.022×10^{23} hydrogen atoms (random)? b) 6.022×10^{23} potassium atoms (random)?

2. What is the mass in grams of: a) 12.044×10^{23} sodium atoms? b) 15.0×10^{23} sodium atoms?

3. Define isotope.

4. Describe the difference between ^{35}Cl and ^{37}Cl.

5. Isotopic abundances are different in other parts of the universe. Suppose that on planet Krypton we find the following stable isotopes and abundances for boron:

   ^{10}B (10.013 amu) 65.75%
   ^{11}B (11.009 amu) 25.55%
   ^{12}B (12.014 amu) 8.70%

   What is the value of the average atomic mass of boron on planet Krypton?

6. Naturally occurring chlorine is composed of ^{35}Cl and ^{37}Cl. The mass of ^{35}Cl is 34.9689 amu and the mass of ^{37}Cl is 36.9659 amu. The average atomic mass of chlorine is 35.453 amu. What are the percentages of ^{35}Cl and ^{37}Cl in naturally occurring chlorine?

7. J. N. Spencer, G. M. Bodner, and L. H. Rickard, *Chemistry: Structure & Dynamics*, Third Edition, John Wiley & Sons, 2006. Chapter 1: Problems: 79, 83, 93.

Model 3: The Mole.

1 dozen items = 12 items
1 mole of items = 6.022×10^{23} items

Critical Thinking Questions

17. a) How many elephants are there in a dozen elephants?

    b) Which has more animals—a dozen elephants or a dozen chickens?

    c) How many elephants are there in a mole of elephants?

    d) Which has more animals—a mole of elephants or a mole of chickens?

    e) Which has more atoms—a dozen H atoms or a dozen Ar atoms?

    f) Which has more atoms—a mole of hydrogen atoms or a mole of argon atoms?

18. Why have scientists chosen to give the number 6.022×10^{23} a name?

19. Which has more atoms: 1.008 g of hydrogen or 39.95 g of argon?

Exercises

8. Without doing any calculations, what is the mass, in grams, of: a) one mole of helium atoms? b) one mole of potassium atoms?

9. What is the average mass, in grams, of: a) one helium atom? b) one potassium atom?

10. What is the mass, in grams, of 5.000 moles of carbon atoms?

11. How many sodium atoms are there in 6.000 moles of sodium?

12. How many sodium atoms are there in 100.0 g of sodium?

13. Calculate the number of <u>atoms</u> in each of the following: a) 50.7 g of hydrogen; b) 1.00 milligram of cobalt; c) 1.00 kilogram of sulfur; d) 1.00 ton of iron.

14. Which element contains atoms that have an average mass of 5.14×10^{-23} grams?

15. What mass of iodine contains the same number of atoms as 25.0 grams of chlorine?

16. J. N. Spencer, G. M. Bodner, and L. H. Rickard, *Chemistry: Structure & Dynamics*, Third Edition, John Wiley & Sons, 2006. Chapter 1: Problems: 89, 95, 100, 103, 105, 113.

Problems

1. Neon has two isotopes with significant natural abundance. One of them, ^{20}Ne, has an atomic mass of 19.9924 amu, and its abundance is 90.5%. Show that the other isotope is ^{22}Ne. Explain your reasoning and include any assumptions that you make.

2. Indicate whether each of the following statements is true or false and explain your reasoning.

   a) On average, one Li atom weighs 6.941 grams.
   b) Every H atom weighs 1.008 amu.
   c) A certain mass of solid Na contains fewer atoms than the same mass of gaseous Ne.
   d) The average atomic mass of an unknown monatomic gas is 0.045 g/mol.

3. The entry in the periodic table for chlorine contains the symbol Cl and two numbers: 17 and 35.453. Give four pieces of information about the element chlorine which can be determined from these numbers (two pieces for each number).

4. The atomic mass of rhenium is 186.2. Given that 37.1% of natural rhenium is rhenium-185, what is the other stable isotope?

ChemActivity 3

Coulombic Potential Energy

(What Is Attractive about Chemistry?)

Model 1: Two Charged Particles Separated by a Distance "*d*".

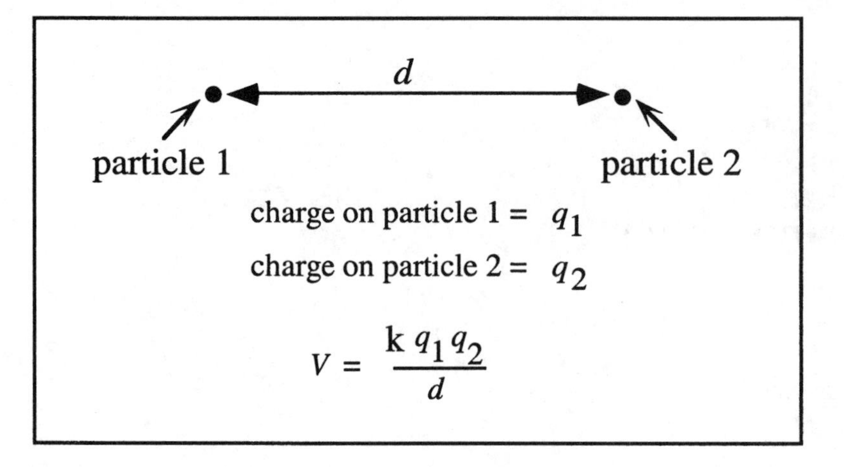

According to Coulomb, the potential energy (V) of two stationary charged particles is given by the equation above, where q_1 and q_2 are the charges on the particles (for example: –1 for an electron), d is the separation of the particles (in pm), and k is a positive-valued proportionality constant.

$$1 \text{ pm} = 10^{-12} \text{ m}$$

Critical Thinking Questions

1. Assuming that q_1 and q_2 remain constant, what happens to the magnitude of V if the separation, d, is increased?

2. If the two particles are separated by an infinite distance (that is, $d = \infty$), what is the value of V?

3. If d is finite, and the particles have the same charge (that is, $q_1 = q_2$), is $V > 0$ or is $V < 0$?

4. If q for an electron is –1,

   a) what is q for a proton?

   b) what is q for a neutron?

   c) what is q for the nucleus of a C atom?

5. Recall that a ^{1}H atom consists of a proton as the nucleus and an electron outside of the nucleus. Is the potential energy of a hydrogen atom a positive or negative number?

Model 2: Ionization Energy.

The ionization energy (IE) is the amount of energy needed to remove an electron from an atom and move it infinitely far away.

Figure 1. Schematic ionization and ionization energies of several hypothetical atoms, each with one proton and one *stationary* electron separated by distance "*d*".

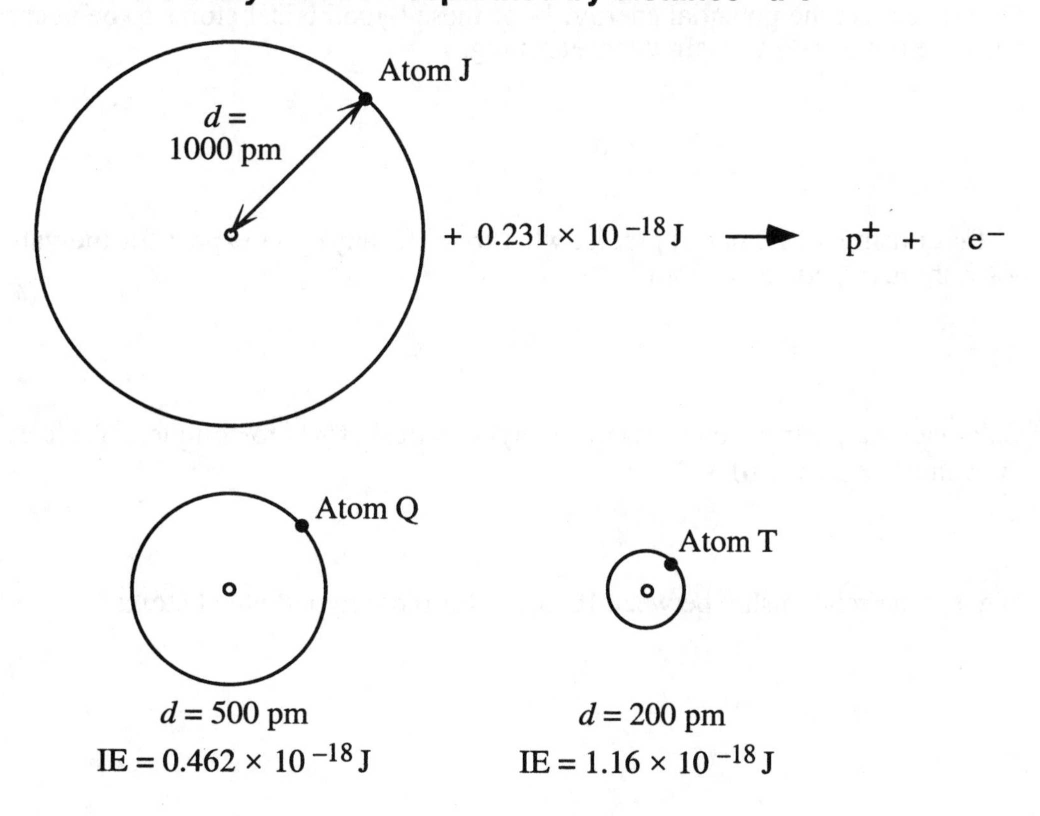

Table 1. Ionization energies of several hypothetical atoms, each with one proton and one *stationary* electron separated by distance "*d*".

Hypothetical Atom	d (pm)	IE (10^{-18} J)	V (10^{-18} J)
A	∞	0	
E	5000.	0.0462	
J	1000.	0.231	
Q	500.0	0.462	
T	200.0	1.16	
Z	100.0	2.31	

Critical Thinking Questions

6. Do you expect the potential energy, V, of these hypothetical atoms to be positive or negative numbers? Explain your reasoning.

7. Without using a calculator, predict what trend (if any) you expect for the values of V for these hypothetical atoms.

8. Calculate the potential energies of the hypothetical atoms to complete Table 1. Use the value $k = 2.31 \times 10^{-16}$ J·pm.

9. What is the relationship between IE and V for these hypothetical atoms?

10. Which of the following systems will have the larger ionization energy? Explain your reasoning.
    a) an electron at a distance of 500 pm from a nucleus with charge +2
    b) an electron at a distance of 700 pm from a nucleus with charge +2

11. Which of the following systems will have the larger ionization energy? Explain your reasoning.
    a) an electron at a distance d_1 from a nucleus with charge +2
    b) an electron at a distance d_1 from a nucleus with charge +1

12. How many times larger is the larger of the two ionization energies from CTQ 11? Show your work.

13. Consider a hydrogen atom and a helium ion, He^+. Which of these do you expect to have the larger ionization energy? Explain your reasoning, including any assumptions you make.

Exercises

1. For a hypothetical atom (as in Table 1) with $V = -5.47 \times 10^{-18}$ J, what would the IE be?

2. Which of the following systems will have the larger ionization energy? Show your work.
    a) an electron at a distance d_1 from a nucleus with charge +2
    b) an electron at a distance $2d_1$ from a nucleus with charge +1

3. Which of the following systems has the larger ionization energy?
    a) an electron at a distance $5d_1$ from a nucleus with a charge of +6
    b) an electron at a distance $6d_1$ from a nucleus with a charge of +7

4. J. N. Spencer, G. M. Bodner, and L. H. Rickard, *Chemistry: Structure & Dynamics*, Third Edition, John Wiley & Sons, 2006. Chapter 3: Problems: 32, 34.

Problems

1. According to the Coulombic Potential Energy equation, if a particle with a charge of –1 is <u>extremely</u> close to a particle with a charge of +2, the potential energy is: a) large and positive b) large and negative c) small and negative d) small and positive

2. Two electrons and one helium nucleus are arranged in a straight line as shown below. The electron on the left is 300 nm from the nucleus; the electron on the right is 400 nm from the nucleus. Write the three Coulombic Potential Energy terms for this arrangement of charges.

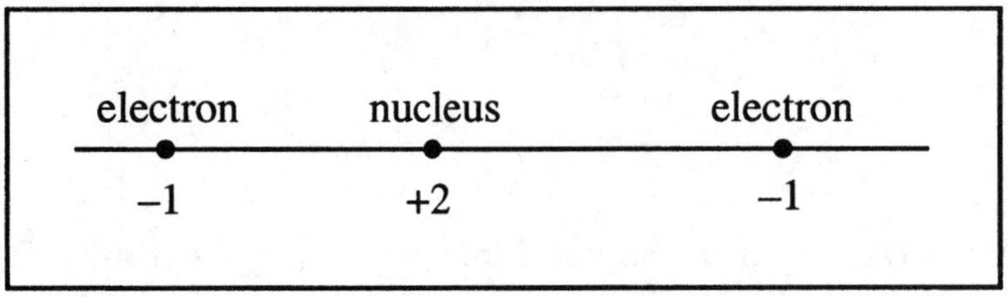

electron	nucleus	electron
−1	+2	−1

ChemActivity 4

The Shell Model (I)

(How Are Electrons Arranged?)

Electrons in atoms are attracted to the nucleus by a Coulombic force. Thus, energy must be supplied (by some means) if the electron is to be pulled away from the nucleus, thereby creating a positively charged species, or cation, and a free electron. For real atoms, the **ionization energy** (IE) of an element is the minimum energy required to remove an electron from a gaseous atom of that element.

Ionization energies are usually obtained experimentally. One method of measuring ionization energies is the electron impact method. Atoms are bombarded with fast-moving electrons. If these electrons have sufficient energy, they will, on colliding with an atom, eject one of the atom's electrons. The ionization energy described above (often called the *first ionization energy*) corresponds to the smallest amount of energy that a bombarding electron needs to be able to knock off one of the atom's electrons.

Model 1: First Ionization Energy (IE_1).

$$M(g) \rightarrow M^{+}(g) + e^{-}$$

For a H atom, IE = 2.178×10^{-18} J.

The first ionization energy, IE_1, for a single atom is a very small number of joules. For reasons of convenience, chemists have chosen to report the ionization energies of elements in terms of the minimum energy necessary to remove a single electron from each of a mole of atoms of a given element. This results in ionization energies for the elements which are in the range of MJ/mole. (Recall that 1 MJ = 10^6 J.)

Critical Thinking Questions

1. How much total energy would it take to remove the electrons from a mole of H atoms? Write this energy in MJ/mole.

2. In CA 1, the electrons were distributed around the nucleus at various distances.

   a) Is the ionization energy of all electrons in the atom the same?

   b) If not, which electron would have the lowest ionization energy (the electron that is closest to the nucleus or the electron that is farthest from the nucleus)?

3. Predict the relationship between IE_1 and atomic number by making a rough graph of IE_1 vs. atomic number. DO NOT PROCEED TO THE NEXT PAGE UNTIL YOU HAVE COMPLETED THIS GRAPH.

Information

Based on our previous examination of ionization energies, it is expected that the ionization energy of an atom would increase as the nuclear charge, Z, increases. In addition, the ionization energy of an atom should decrease if the electron being removed is moved farther from the nucleus (that is, if *d* increases).

Table 1 below presents the experimentally measured ionization energies of the first 20 elements. We will examine these results and attempt to propose a model for the structure of atoms based on these data.

Table 1. First Ionization energies of the first 20 elements.

Z		IE_1 (MJ/mole)	Z		IE_1 (MJ/mole)
1	H	1.31	11	Na	0.50
2	He	2.37	12	Mg	0.74
3	Li	0.52	13	Al	0.58
4	Be	0.90	14	Si	0.79
5	B	0.80	15	P	1.01
6	C	1.09	16	S	1.00
7	N	1.40	17	Cl	1.25
8	O	1.31	18	Ar	1.52
9	F	1.68	19	K	0.42
10	Ne	2.08	20	Ca	0.59

Critical Thinking Questions

4. Compare your answer to CTQ 3 to the data in Table 1. Comment on any similarities and differences.

5. Using grammatically correct English sentences:

   a) provide a possible explanation for why IE_1 for He is greater than IE_1 for H.

   b) provide a possible explanation for why IE_1 for Li is less than IE_1 for He.

Model 2: Shell Model Diagrams for Hydrogen and Helium Atoms.

One model of the hydrogen atom, often referred to as the Bohr model, pictures the H atom as a nucleus of charge +1 surrounded by an electron in an orbit of some distance, as shown in Figure 1.

Figure 1. Schematic diagram of a hydrogen atom based on the Bohr model.

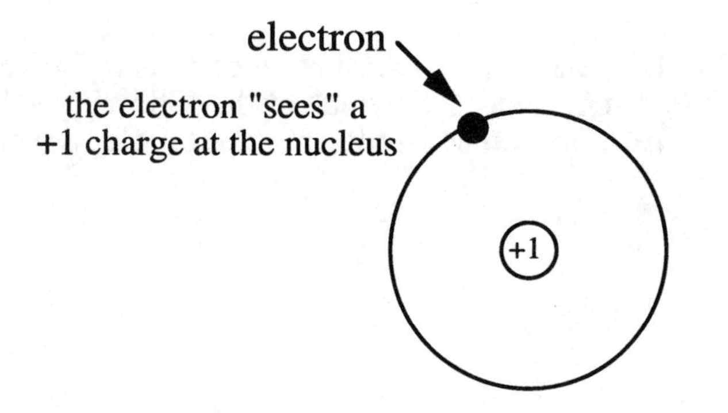

Examining the data in Table 1, we note that the ionization energy of He (Z=2) is larger than that of H (Z=1) by approximately a factor of 2. This is consistent with the two electrons in the He atom orbiting the He nucleus at a distance approximately the same as that in H, as shown in Figure 2.

Figure 2. Diagram of a helium atom using the shell model.

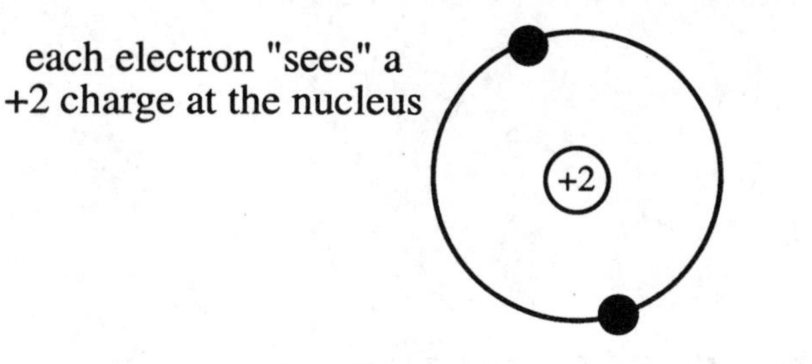

Critical Thinking Question

6. The value of the ionization energy of He given in Table 1 is described as being consistent with two electrons in a "shell" approximately the same distance from the nucleus as the one electron in H. Explain how this conclusion can be reached.

Information

Because the He nucleus has a charge of +2, we would expect that the ionization energy to remove an electron from (approximately) the same distance as in a H atom would be (approximately) twice that of the H atom. That is what we observe. We can say that there are two electrons in a **shell** around the He nucleus. Although we will present figures in which the shells appear to be circular (mostly because it is difficult to present three-dimensional representations on paper), we recognize that the model we develop is qualitatively consistent with spherical shells. Thus, within our *Shell Model*, He consists of a nucleus surrounded by 2 electrons in a single shell.

7. Recall that the IE of H is 1.31 MJ/mole. If all three electrons in Li were in the first shell at a distance equal to that of hydrogen, which of the following values would be the better estimate of the IE_1 of Li: 3.6 MJ/mole or 0.6 MJ/mole? Explain.

8. Why is the IE_1 (in Table 1) for Li inconsistent with placing a third electron in the first shell at a distance approximately equal to that of the electron in H?

Model 3: The Shell Model for Lithium.

For Li, there is a change in the trend of the ionization energy. The ionization energy of a Li atom is *less* than that of He. In fact, it is significantly *smaller* than that of the H atom! This is not consistent with a model of placing a third electron in the first shell, for doing so would result in an ionization energy which is larger than that of He. In order for Li to have a lower ionization energy than H, either the nuclear charge Z must be lower than that of H, or the distance of the easiest-to-remove electron from the nucleus must be greater than in H (and He), or both. We know that the nuclear charge is *not* lower than that of H; thus, the electron being removed must be farther from the nucleus than the first shell. Although the data we have does not require us to choose the following model, let us assume that the structure of Li involves two electrons in a first shell (as in He) with the third electron placed in a second shell, with a significantly larger radius, as shown in Figure 3.

Figure 3. Diagram of a lithium atom using the shell model.

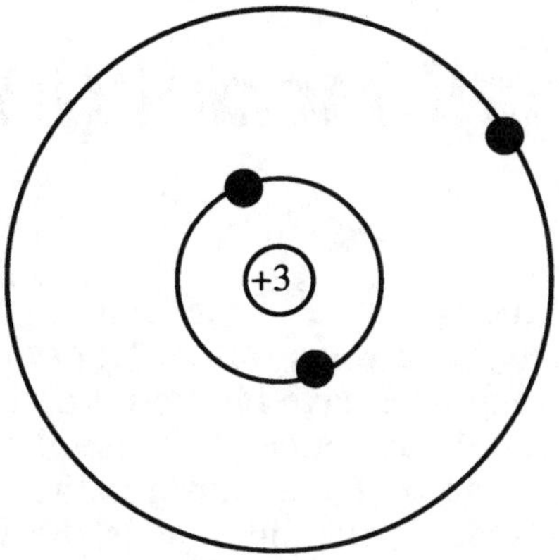

+3

Critical Thinking Question

9. How is the model presented in Figure 3 consistent with the data in Table 1?

Exercises

1. A scientist proposes a model for the helium atom in which both electrons are in a "shell" which is half the distance from the nucleus as the electron in a hydrogen atom. Is this model consistent with the data in Table 1? Explain your reasoning. (Hint: according to the Coulombic Potential Energy equation, how much more strongly does a nuclear charge of +2, as in He, hold an electron than a nuclear charge of +1, as in H? According to the Coulombic Potential Energy equation, how much more strongly does a nuclear charge hold an electron if it is at $d/2$, rather than d?)

2. Propose an alternative model for the lithium atom which is consistent with the data in Table 1.

Problem

1. a) Write the three Coulombic Potential Energy terms for the helium atom model in Figure 2. Assume that the distance between each electron and the nucleus is d and that the distance between the two electrons is $2d$. b) Based on your answer to part a) explain why the IE of He is slightly less than twice the IE of H even though both atoms are about the same size.

ChemActivity **5**

The Shell Model (II)

Model 1: Core Charge.

Notice that within this model of the structure of the Li atom, the outermost electron (which is the easiest-to-remove electron) is always farther from the nucleus than the two inner electrons in the first shell. Although we have ignored it up to this point, we should remember that all of the electrons repel each other because they are each negatively charged. (Recall that the negative charge on the electron is the same magnitude as the positive charge on each proton.) Of particular interest is the repulsion of the outer electron by the two inner electrons. This dramatically decreases the overall force of attraction pulling the outer electron toward the nucleus. Thus, the net charge acting on the outer electron (to hold on to it) is the charge of the inner core of the atom, consisting of the nucleus and the two inner-shell electrons, which is (+3) + (–2) = +1. The nucleus plus the inner shells of electrons constitute the **core** of the atom, and the net overall charge on the core is the **core charge**. The electrons in the outermost shell of the atom are referred to as **valence** electrons. Thus, H has one valence electron and no inner-shell electrons; Li has one valence electron and two inner-shell electrons. Note that the core charge is a positive number. We can represent the Li atom in terms of core charge as shown in Figure 1.

Figure 1. Diagram of a lithium atom using the shell model (a) and the core charge concept (b).

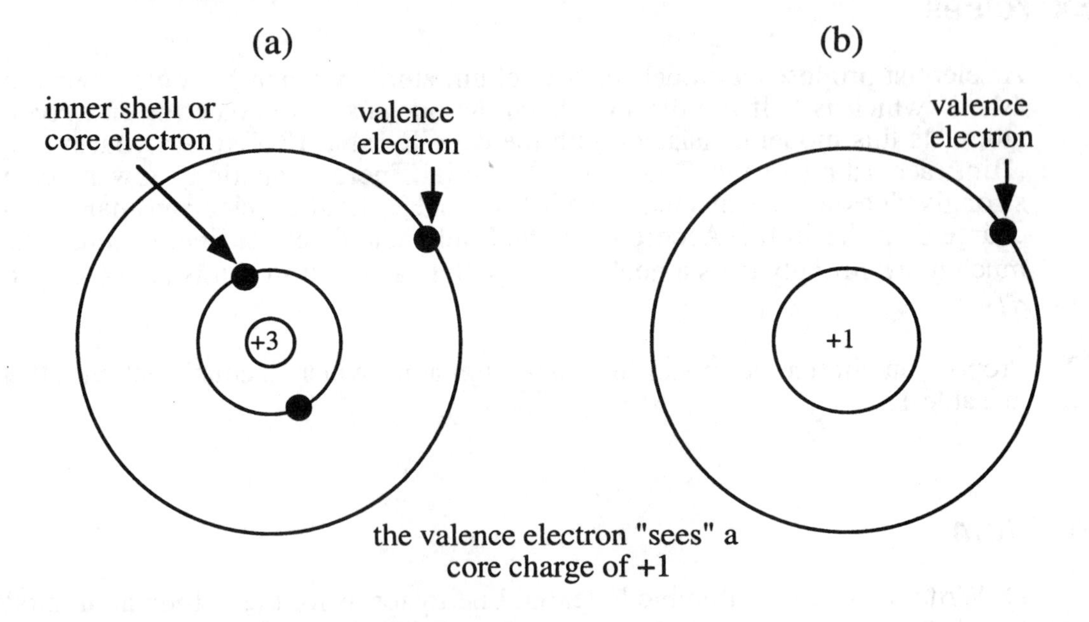

Critical Thinking Question

1. a) How many electrons are in the valence shell of H? Of He? Of Li?

   b) How many inner shell (core) electrons does H have? He? Li?

   c) What is the core charge of H? Of He? Of Li?

Model 2: The Beryllium Atom.

Figure 2. Diagram of a Be atom using the shell model (a) and the core charge concept (b).

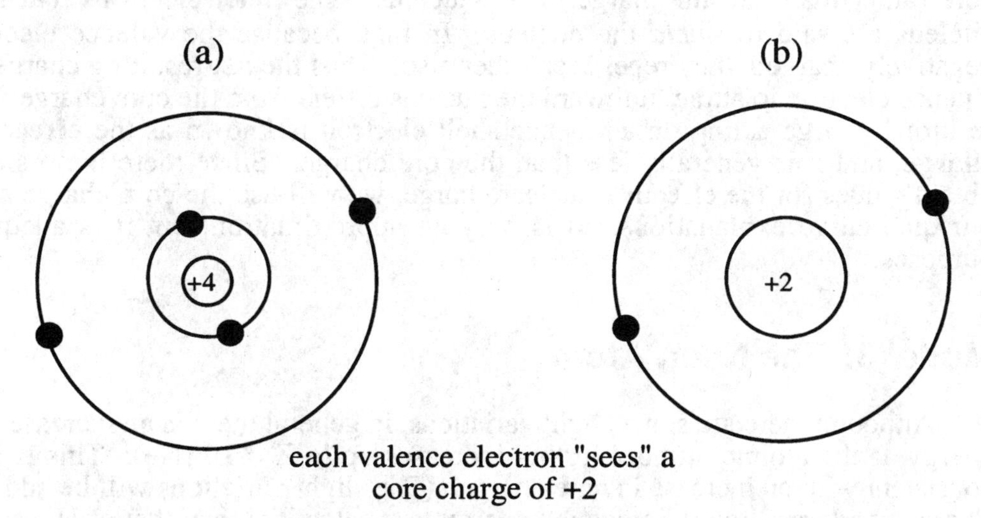

The next element, Be, has an ionization energy which is larger than that for Li. This is consistent with the fourth electron in Be being added to the second shell. Thus, Be has 2 valence electrons and a core charge of +2. Two representations of the Be atom are given in Figure 2.

Critical Thinking Questions

2. a) Why is the nuclear charge of Be "+4"?

   b) How many inner shell (core) electrons does Be have?

   c) How many valence electrons does Be have?

   d) Show how the core charge for Be was calculated.

   e) What is the relationship between the number of valence electrons and the core charge of a neutral atom?

3. Explain how the core charges of Li and Be are consistent with the IE_1 values for these two atoms in Table 1 of **ChemActivity 4: Shell Model (I)**.

Information

As described above, the outer shell valence electrons experience the charge of the core rather than the full charge of the nucleus. The inner electrons that surround the nucleus are said to *shield* the nucleus. In fact, because the valence electrons are all negatively charged, they repel each other also. Thus the net resulting charge acting on a valence electron to attract it toward the nucleus differs from the core charge. This overall resulting charge acting on a valence shell electron is known as the **effective nuclear charge**, and it is generally less than the core charge. Since there is no simple way to obtain values for the effective nuclear charge, we will use the core charge as a basis for our qualitative explanations. It is only an approximation, but it is adequate for our purposes.

Model 3: The Neon Atom.

Although there are some slight variations, in general there is an increase in ionization energy as the atomic number further increases up to Z = 10 (Ne). This is qualitatively consistent with an increase in core charge. (The slight variations will be addressed later.) There is no large drop in ionization energy to a value less than that of H, as we observed in going from He to Li, to indicate that a third shell is needed. This suggests that as we move from Be up to Ne, the number of electrons in the second shell increases.

Figure 3. Diagram of a Ne atom using the shell model (a) and the core charge concept (b).

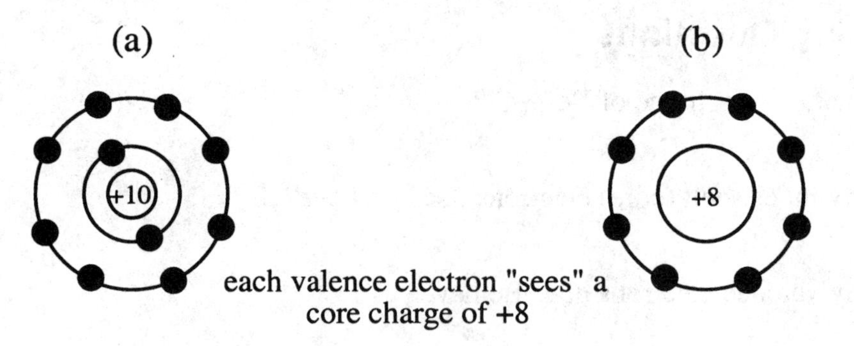

Ne has 8 electrons in the second (valence) shell, and 2 electrons in the inner (first) shell. Notice that we can number the shells based on their distance from the nucleus. We can let the number "n" represent the number of the shell an electron is in. Thus, Ne has 2 electrons in the $n = 1$ shell and 8 electrons in the $n = 2$ shell.

Critical Thinking Questions

4. Show how the core charge for Ne was calculated.

5. Make a diagram, similar to those in Figure 3, for the nitrogen atom.

Model 4: The Sodium Atom.

In moving to atomic number 11, Na, we observe a dramatic drop in ionization energy from that of Ne (Z=10). This decrease is analogous to (and similar in magnitude to) that observed in going from He to Li. Note that the ionization energy of Na is only 0.50 MJ/mole, even less (although only slightly so) than that of Li. Analogous to the conclusions we reached concerning the structure of the Li atom, these results suggest that the eleventh electron in Na should be placed in a third shell (n = 3), at a slightly greater distance from the nucleus than the second shell is for Li. Thus, it appears that the n = 2 shell can accommodate only eight electrons. (Recall that the n = 1 shell holds only two.) The resulting model for the Na atom is shown in Figure 4.

Figure 4. Diagram of a Na atom using the shell model (a) and the core charge concept (b).

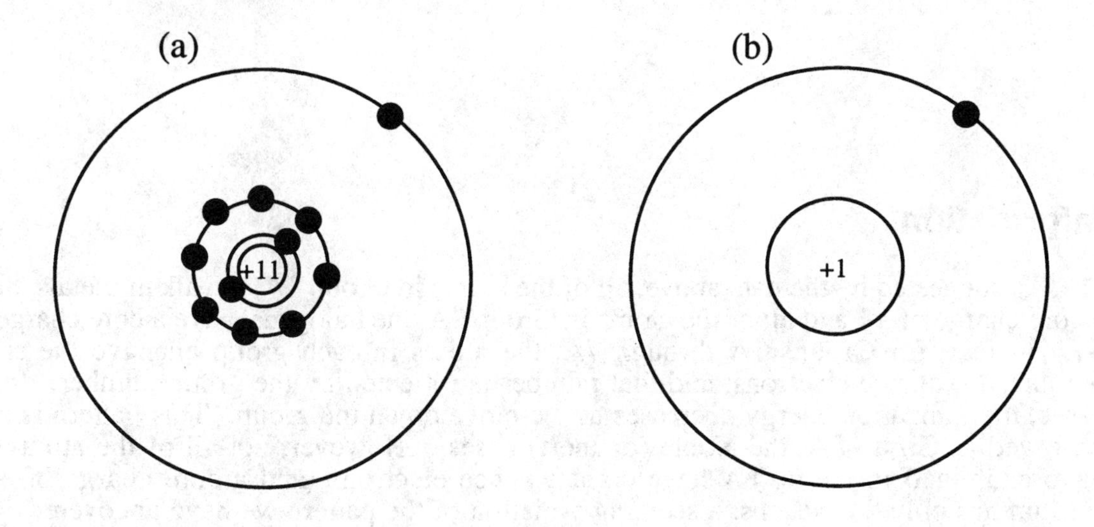

The pattern of ionization energies for the elements with Z = 11 to Z = 18 follows the trend we previously identified for Z = 3 to Z = 10: a general increase (with slight variations).

Critical Thinking Questions

6. How many electrons does Na have in shell $n = 1$? $n = 2$? $n = 3$?

7. Show how the core charge for Na was calculated.

8. Based on ionization energy data, Table 1 of **ChemActivity 4: Shell Model (I)**, is the radius of the valence shell of Na larger, smaller or the same as the radius of the valence shell of Li?

9. Explain how the core charges of Na and Ne are consistent with the IE_1 data in Table 1 of **ChemActivity 4: Shell Model (I)**.

10. Why are the IE_1 data *not* consistent with Na having nine electrons in the second shell?

Information

As suggested by the data above, all of the atoms in Group 1A, the alkali metals, have a core charge of +1 and all of the atoms in Group 7A, the halogens, have a core charge of +7. In fact, for Groups 1A through 7A, the atoms in each group all have the same number of valence electrons, and that number is reflected by the group number. In all cases, the ionization energy decreases as we move down the group. This pattern is also observed in Group 8A, the Noble (or Inert) gases . However, not all of the atoms we have examined in Group 8A have eight valence electrons (and a core charge of +8). Helium has only 2 electrons, a seeming violation of the pattern we have uncovered. The resolution of this apparent inconsistency is that although He has only 2 valence electrons, its valence shell is *completely filled*. The same is true of Ne, although for Ne a filled valence shell has 8 electrons. Thus, we find that the structure of the elements using this Shell model is reflected in the placement of the elements in the periodic table.

Model 5: The Shell Model and Ionization Energies.

Table 1. Atomic Properties of Various Atoms.

Element	Valence Shell (n)	Number of Valence Electrons	Core Charge	IE_1 (MJ/mole)
H	1	1	+1	1.31
Li	2	1	+1	0.52
Na	3	1	+1	0.50
Rb				0.40
F	2	7	+7	1.68
Cl	3	7	+7	1.25

Critical Thinking Questions

11. Locate H, Li, and Na on the periodic table.

a) Describe any relationship between the core charge of these atoms, the number of valence electrons, and their positions in the periodic table.

b) Describe any relationship between the valence shell of these atoms and their positions in the periodic table.

c) Based on its position in the periodic table, predict the valence shell, core charge, and number of valence electrons for Rb.

d) Using the shell model and referring to the Coulombic Potential Energy relationship (equation 1, CA3), explain clearly how the IE_1 for Rb is consistent with your answer to part c.

12. Construct a shell model diagram of F that is consistent with the information in Table 1.

13. Locate F and Cl on the periodic table.

    a) Describe any relationship between the core charge of these atoms, the number of valence electrons, and their position in the periodic table.

    b) Describe any relationship between the valence shell of these atoms and their position in the periodic table.

    c) Within our model and referring to the Coulombic Potential Energy expression, explain why the IE_1 of Cl is less than that of F.

14. Based on its position in the periodic table, what is the valence shell and what is the core charge for C. Explain your reasoning.

15. How does the core charge on the neutral atom change as we move from left to right across a row (period) of the periodic table?

16. Within our model and referring to the Coulombic Potential Energy expression, explain why the IE increases from left to right across a row of the periodic table.

Exercises

1. How many valence electrons are there in: a) C? b) O? c) N? d) Ne?

2. What is the core charge for: a) C? b) O? c) N? d) Ne?

3. Based on the information in Table 1 of **ChemActivity 4: Shell Model (I)**, estimate the ionization energy for Br. Explain your reasoning.

4. If a single electron is removed from a Li atom, the resulting Li^+ cation has only two electrons, both in the $n = 1$ shell. In this respect it is very similar to a He atom. How would you expect the ionization energy of a Li^+ cation to compare to that of a He atom? Explain your reasoning.

5. If a single electron is somehow added to a F atom, the resulting F^- anion has a total of 8 valence electrons in the $n = 2$ shell. In this respect it is very similar to a Ne atom. How would you expect the ionization energy of a F^- anion to compare to that of a Ne atom? Explain your reasoning.

6. Predict the order of the ionization energies for the atoms Br, Kr, and Rb. Explain your reasoning.

7. The radius of the outer shell in Li is larger than the radius of the inner shell. Which electron is harder to remove—the valence electron or one of the inner shell electrons? Explain.

8. Without referring to a table of ionization energies, J. N. Spencer, G. M. Bodner, and L. H. Rickard, *Chemistry: Structure & Dynamics*, Third Edition, John Wiley & Sons, 2006. Chapter 3: Problems: 43, 44, 58, 61, 64, 65, 204.

Problems

1. Indicate whether each of the following statements is TRUE or FALSE and explain your reasoning.

   a) The core charge of Br is +7.
   b) Helium has the largest 1st ionization energy.

2. Explain how the model of the structure of Be having the fourth electron in a third shell, further from the nucleus than any of the three electrons in Li, is *not* consistent with the experimentally obtained ionization energies.

ChemActivity 6

Atomic Size

(What Size Are Atoms?)

Model 1: Atomic Properties and the periodic table.

> Ionization energies increase as the core charge increases across a row (period) of the periodic table.
>
> Ionization energies increase when the core charge remains the same and the valence electrons are in a shell closer to the nucleus (that is, up a column (group) of the periodic table).

We have seen that there are trends in a physical property, IE_1, related to the position of the elements in the periodic table. Many other physical and chemical properties of the elements in a particular group, such as the alkali metals or the halogens, are relatively similar. This suggests that many properties of an atom are related to the number of valence electrons present. The nuclear charge (and core charge) can also be important in determining atomic characteristics because this determines the strength of attraction between the nucleus and the valence electrons. This was the basis for our previous analysis of first ionization energies.

One measure of the size of an atom (or ion) is the covalent radius of the atom. This can be thought of as the radius of the outermost shell of an atom or ion.

Table 1. Atomic radii of various atoms.

Element	Valence Shell (n)	Core Charge	Radius (pm)
Boron	2	+3	89
Carbon	2	+4	77
Oxygen	2	+6	66
Sulfur	3	+6	104
Arsenic	4	+5	121
Selenium	4	+6	117

Critical Thinking Questions

1. What is the relationship between the valence shell of each atom in Table 1 and its position in the periodic table?

2. Why does the core charge increase as one moves from left to right across a period in the periodic table—for example, from boron to carbon to oxygen?

3. What trend in atomic radius is observed as one moves from left to right across a period? Why?

4. What trend in atomic radius is observed as one moves down a group in the periodic table? Why?

5. Estimate the radii of three atoms not listed in Table 1, based on the data presented. Explain how you are able to estimate these values from the data given.

Table 2. Ionic radii of various isoelectronic ions.

Ion	Valence Shell (n)	Core Charge	Radius (pm)
S^{2-}	3	+6	184
Cl^{-}	3	+7	181
K^{+}	3	+9	133
Ca^{2+}	3	+10	104

Chemical species that have identical numbers of electrons are **isoelectronic**.

6. How many electrons do the ions in Table 2 have?

7. Provide a shell model diagram for K^{+} showing all electrons explicitly and then show how the core charge for K^{+} was calculated.

8. What is the basis for the trend in ionic radii seen in Table 2?

9. Predict which is larger: the O^{2-} ion or the F^{-} ion. Explain.

Exercises

1. Based on the data in Tables 1 and 2, estimate the radius of each of the following species. Explain your reasoning.
   a) Ar b) N c) F^- d) Ne

2. Indicate whether each of the following statements is true or false and explain your reasoning.

   a) The first ionization energy of Ba is expected to be larger than that of Mg.
   b) The Na^+ ion is expected to have a larger radius than a Ne atom.
   c) The radius of Cl^- is expected to be larger than the radius of Ar.
   d) The radius of Ar is expected to be larger than that of Ne.
   e) The first ionization energy of Ar is expected to be greater than that of Ca^{2+}.

3. J. N. Spencer, G. M. Bodner, and L. H. Rickard, *Chemistry: Structure & Dynamics*, Third Edition, John Wiley & Sons, 2006. Chapter 3: Problems: 155, 159, 160, 166, 199, 200.

Model 2: The Effect of Additional Electrons on Size.

Table 3. Atomic radii of various atoms and ions.

Atom or Ion	Valence Shell (*n*)	Core Charge	Number of Valence Electrons	Radius (pm)
F	2	+7	7	64
F^-	2	+7	8	133
O	2	+6	6	66
O^{2-}	2	+6	8	140

Electrons repel electrons. Therefore, as extra electrons are added to the valence shell of an atom (or ion), the radius of the atom (or ion) increases. Similarly, as electrons are removed from an atom (or ion), the radius of the atom (or ion) decreases.

Critical Thinking Questions

10. Why do electrons repel each other?

11. What are three characteristics of an atom (or ion) which must be considered in determining its relative radius?

Exercises

4. Which is the smaller species in each of the following groupings?
   a) N , N^{3-}
   b) K , K^+
   c) Cl , Cl^-
   d) H , H^-
   e) Mg , Mg^{2+}

5. Which is larger— Fe^{2+} or Fe^{3+} ?

6. Which is the largest species in each of the following groupings?
   a) Pb , Pb^{2+} , Pb^{4+}
   b) Mg , Al , Na
   c) Mg^{2+} , Ca^{2+} , Ba^{2+} ,
   d) H , H^- , H^+
   e) Na , Cl , Br, I , Rb
   f) P^{3-}, S^{2-}, Cl^-, F^-.

7. J. N. Spencer, G. M. Bodner, and L. H. Rickard, *Chemistry: Structure & Dynamics*, Third Edition, John Wiley & Sons, 2006. Chapter 3: Problems: 164, 165, 167-172, 201.

Problems

1. Which of the following elements should have the largest second ionization energy? Na; Mg; Al; Si; P. Explain.

2. Mg atoms are larger than S atoms. Explain why Mg^{2+} ions are smaller than S^{2-} ions.

ChemActivity 7

Electromagnetic Radiation

Model 1: A Wave and Its Wavelength.

The figure below represents part of a **wave**. The entire wave can be thought of as extending infinitely in both directions. One important characteristic of a wave is its **wavelength** (λ), which is the distance between two consecutive peaks (or troughs) in the wave.

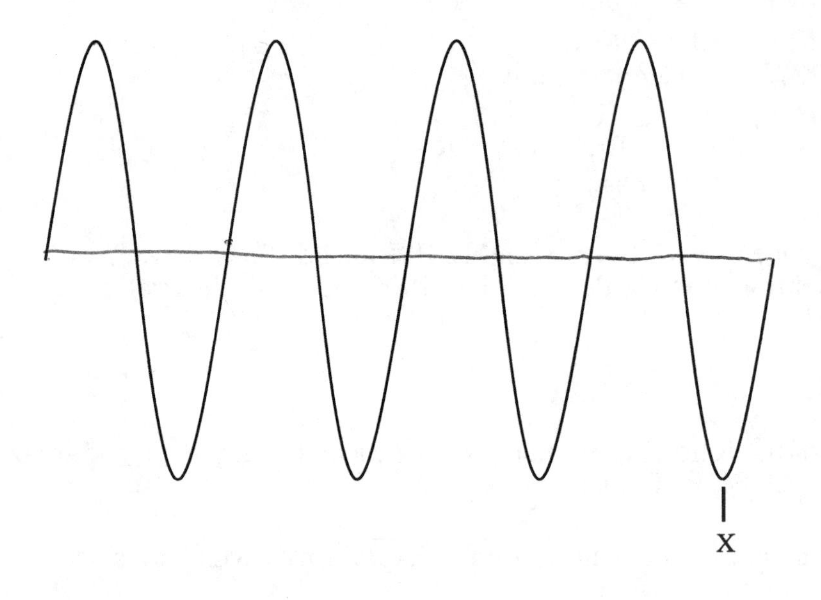

Critical Thinking Questions

1. On the figure above, draw a line connecting two points whose separation is equal to the wavelength of the wave. If there is more than one way to do this, draw a second line.

2. Suppose that the wave depicted above were traveling to the right at a speed of 35 cm/sec, and that λ = 2.5 cm.

   a) How long would it take for 1 wavelength (or 1 cycle of the wave) to travel past the point X?

   b) How many wavelengths (or cycles) would travel past the point X during a time interval of 1 second?

c) Would your answer to part a) increase, decrease, or remain the same if $\lambda >$ 2.5 cm. Explain your reasoning.

d) Would your answer to part b) increase, decrease, or remain the same if $\lambda >$ 2.5 cm. Explain your reasoning.

3. The **frequency** (f) of a wave is defined as the number of wavelengths per second which travel past a given point.

   a) For a wave traveling at a given **speed**, c, how does the frequency depend on the wavelength, if at all?

   b) Provide a mathematical expression showing the relationship between f, λ, and c for a wave. (Hint: consider how you determined answers to CTQ 2ab).

Exercise

1. Indicate whether the following statement is true or false, and explain your reasoning:

   For waves traveling at the same speed, the longer the wavelength the greater the frequency.

Information

Read materials assigned by your instructor on electromagnetic radiation, quantization of energy, and atomic spectra.

Model 2: Electromagnetic Radiation and Photons.

Light can be thought of as an **electromagnetic wave** or **electromagnetic radiation** having a particular wavelength and frequency. In addition, Albert Einstein proposed almost a century ago that electromagnetic radiation can be viewed as a stream of particles known as **photons**, each of which has a particular amount of energy associated with it. Specifically, he proposed the following equation:

$$E_{photon} = hf$$ where h is called Planck's constant.

Table 1. Wavelengths, frequencies, and energies of electromagnetic radiation.

Wavelength (nm)	Frequency (10^{14} s^{-1})	Energy (10^{-19} J)
333.1	9.000	5.963
499.7	6.000	3.976
999.3	3.000	1.988

$1 \text{ nm} = 10^{-9} \text{ m}$

The joule (J) is a unit of energy. $1 \text{ J} = 1 \text{ kg m}^2/\text{s}^2$

Table 2. Regions of the electromagnetic spectrum.

Region	Wavelength Range
radiowave	3 km – 30 cm
microwave	30 cm – 1 mm
infrared (IR)	1 mm – 800 nm
visible (VIS)	800 nm – 400 nm
ultraviolet (UV)	400 nm – 10 nm
X-ray	10 nm – 0.1 nm
gamma ray	< 0.1 nm

Critical Thinking Questions

4. A certain photon has a wavelength of 100 nm. In what region of the electromagnetic spectrum should this photon be classified (see Table 2)?

5. According to the data in Table 1 and the equation proposed by Einstein, what is the value of Planck's constant (include the units)?

6. Based on the data in Table 1 and the relationship between speed, frequency, and wavelength, what is the speed of electromagnetic radiation (light waves)?

Information

Quantities a and b are **proportional** when $a = kb$, where k is some constant. The two quantities are said to be **inversely proportional** when $a = k/b$.

Critical Thinking Questions

7. Write the mathematical equation that relates the energy of a photon and its frequency. Is the energy of a photon proportional or inversely proportional to *f*?

8. Write the mathematical equation that relates the energy of a photon and its wavelength. Is the energy of a photon proportional or inversely proportional to λ?

Exercises

2. Complete the following table:

Energy (J)	Wavelength (m)	Frequency (s^{-1})	Region of Spectrum
9.939×10^{-20}	1.999×10^{-6}	1.50×10^{14}	
3.975×10^{-19}	0.500×10^{-6}	5.9996×10^{14}	
9.94×10^{-19}			
	1.00×10^{-9}		

3. Which is the more energetic, a red photon (λ ~ 700 nm) or a blue photon (λ ~ 400 nm)? Explain.

4. J. N. Spencer, G. M. Bodner, and L. H. Rickard, *Chemistry: Structure & Dynamics*, Third Edition, John Wiley & Sons, 2006. Chapter 3: Problems: 7-10.

Problem

1. The first ionization energy of Na(g) is 0.50 MJ/mole. Can a photon with a wavelength of 500 nm ionize a sodium atom? Explain.

ChemActivity **8**

Photoelectron Spectroscopy

(What Is Photoelectron Spectroscopy?)

From our previous examination of the ionization energies of the atoms, we proposed a shell model of the atom, and noted that the number of valence electrons in the outermost shell is related to the position of the element in the periodic table, and therefore is an important factor in determining the physical and chemical properties of the element. Within this model, the electrons in an atom are arranged in shells about the nucleus, with the successive shells being farther and farther from the nucleus. The ionization energy described previously is the minimum energy needed to remove an electron from the atom. The most easily removed electron always resides in the valence shell, since that is the shell that is the farthest from the nucleus. For atoms with many electrons, we would expect that the energy needed to remove an electron from an inner shell would be greater than that needed to remove an electron from the valence shell, because an inner shell is closer to the nucleus and is not as fully shielded as the outer valence electrons. Thus, less energy is needed to remove an electron from an $n = 2$ shell than from an $n = 1$ shell, and even less is needed to remove an electron from an $n = 3$ shell. But do all electrons in a given shell require precisely the same energy to be removed? In order to answer this question, we must consider ionization energies in greater detail.

Model 1: Ionization Energies and Energy Levels

From the Coulombic Potential Energy expression, we know that an electron in a given shell will require a certain energy to be separated from the atom. Thus, an electron can be said to occupy an **energy level** in an atom. Within our model, each electron must be in a shell at a particular distance from the nucleus, and the energy levels corresponding to these shells are **quantized**—that is, only certain discrete energy levels should be found.

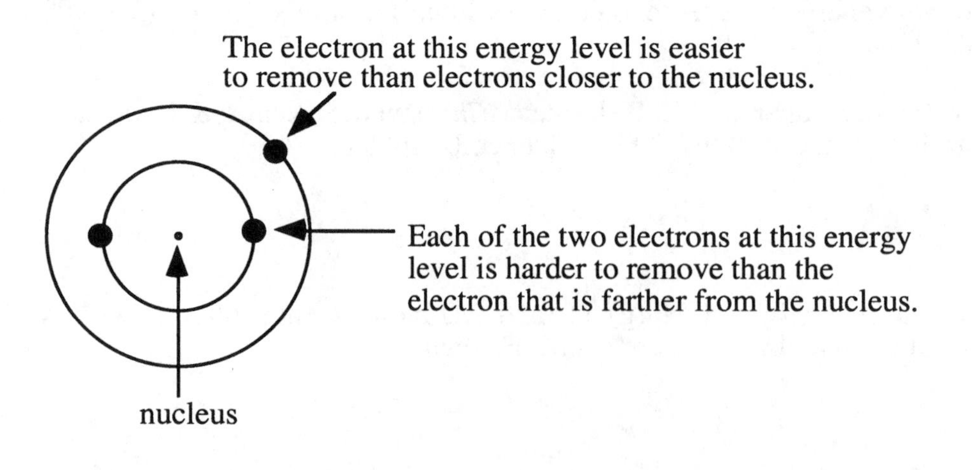
The electron at this energy level is easier to remove than electrons closer to the nucleus.
Each of the two electrons at this energy level is harder to remove than the electron that is farther from the nucleus.
nucleus

Critical Thinking Question

1. Suppose that the values for the two energy levels for the atom in Model 1 are –0.52 MJ/mole and –6.26 MJ/mole.

   a) Which of these two energy levels is the lower energy level?

   b) Clearly identify which energy level is associated with each of the three electrons given in Model 1.

   c) Determine the ionization energies of each of the three electrons given in Model 1.

Model 2: Photoelectron Spectroscopy.

Ionization energies may be measured by the electron impact method, in which atoms in the gas phase are bombarded with fast-moving electrons. These experiments give a value for the ionization energy of the electron that is most easily removed from the atom—in other words, the ionization energy for an electron in the highest occupied energy level. An alternative, and generally more accurate, method that provides information on all the occupied energy levels of an atom (that is, the ionization energies of all electrons in the atom) is known as photoelectron spectroscopy; this method uses a photon (a packet of light energy) to knock an electron out of an atom. Electrons obtained in this way are called photoelectrons.

Very high energy photons, such as very-short-wavelength ultraviolet radiation, or even x-rays, are used in this experiment. The gas phase atoms are irradiated with photons of a particular energy. If the energy of the photon is greater than the energy necessary to remove an electron from the atom, an electron is ejected with the excess energy appearing as kinetic energy, mv^2, where v is the velocity of the ejected electron. In other words, the speed of the ejected electron depends on how much excess energy it has received. So, if IE is the ionization energy of the electron and KE is the kinetic energy with which it leaves the atom, we have

$$E_{photon} = IE + KE$$

or, upon rearranging the equation,

$$IE = E_{photon} - KE$$

Thus, we can find the ionization energy, IE, if we know the energy of the photon and we can measure the kinetic energy of the photoelectron. The kinetic energy of the electrons is measured in a photoelectron spectrometer.

Figure 1: Photoelectron spectroscopy of a hypothetical atom.

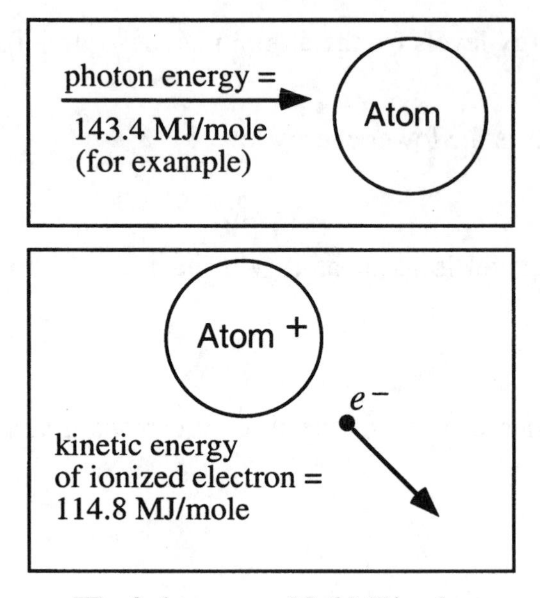

IE of electron = 28.6 MJ/mole

If photons of sufficient energy are used, an electron may be ejected from *any* of the energy levels of an atom. Each atom will eject only one electron, but every electron in each atom has an (approximately) equal chance of being ejected. Thus, for a large group of identical atoms, the electrons ejected will come from all possible energy levels of the atom. Also, because the photons used all have the same energy, electrons ejected from a given energy level will all have the same energy. Only a few different energies of ejected electrons will be obtained, corresponding to the number of energy levels in the atom.

The results of a photoelectron spectroscopy experiment are conveniently presented in a *photoelectron spectrum.* This is essentially a plot of the number of ejected electrons (along the vertical axis) vs. the corresponding ionization energy for the ejected electrons (along the horizontal axis). It is actually the kinetic energy of the ejected electrons that is measured by the photoelectron spectrometer. However, as shown in the equation above, we can obtain the ionization energies of the electrons in the atom from the kinetic energies of the ejected electrons. Because these ionization energies are of most interest to us, a photoelectron spectrum uses the ionization energy as the horizontal axis.

Figure 2: A simulated photoelectron spectrum of the hypothetical atom in Figure 1.

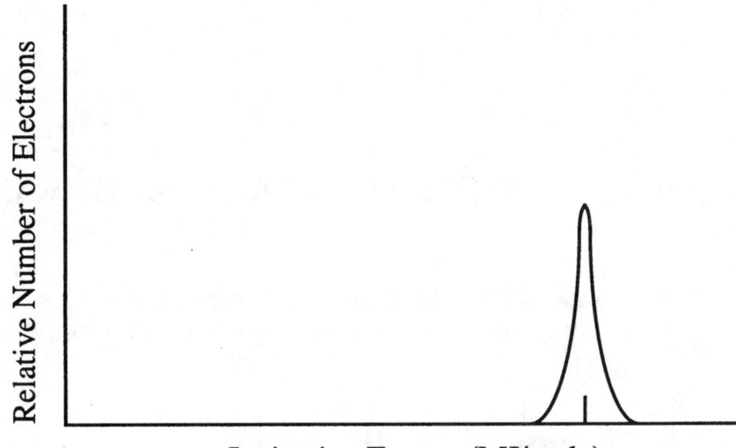

Critical Thinking Questions

2. Use the data presented in Figure 1 to verify that the IE of the ejected electron is 28.6 MJ/mole.

3. What determines the position of each peak (where along the horizontal axis the peak is positioned) in a photoelectron spectrum?

4. What is the numerical value at the position of the hatch mark in the photoelectron spectrum of Figure 2?

5. What is meant by the term "energy level"?

6. What is the value of the energy level of the electron in Figure 2?

7. What determines the height (or intensity) of each peak in a photoelectron spectrum?

8. Explain why it is not possible to determine the number of electrons in an individual hypothetical atom from the photoelectron spectrum in Figure 2.

Model 3: The Energy Level Diagram of Another Hypothetical Atom.

A *hypothetical* atom in a galaxy far, far away has 2 electrons at one energy level and 3 electrons at another energy level as shown in the energy level diagram below:

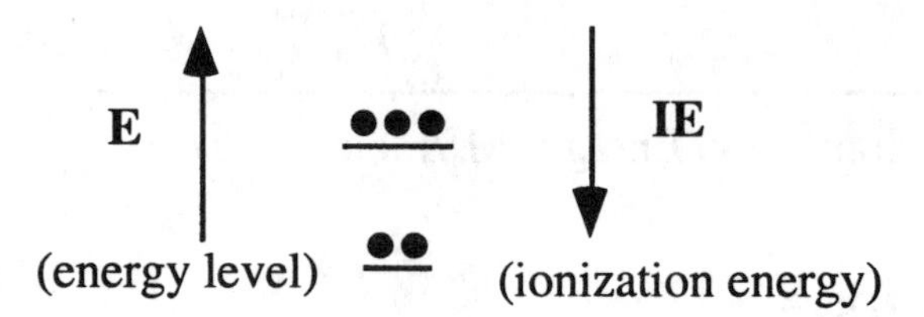

Critical Thinking Questions

9. How many peaks (1,2,3,4,5) will appear in a photoelectron spectrum of a sample of this hypothetical atom? Why?

10. Describe the relative height of the peaks in the photoelectron spectrum of a sample of this hypothetical atom.

11. Suppose that the two energy levels are –0.85 MJ/mole and –4.25 MJ/mole. On the axes below, make a sketch of the photoelectron spectrum of a sample of this hypothetical atom. Make sure to label the axes appropriately.

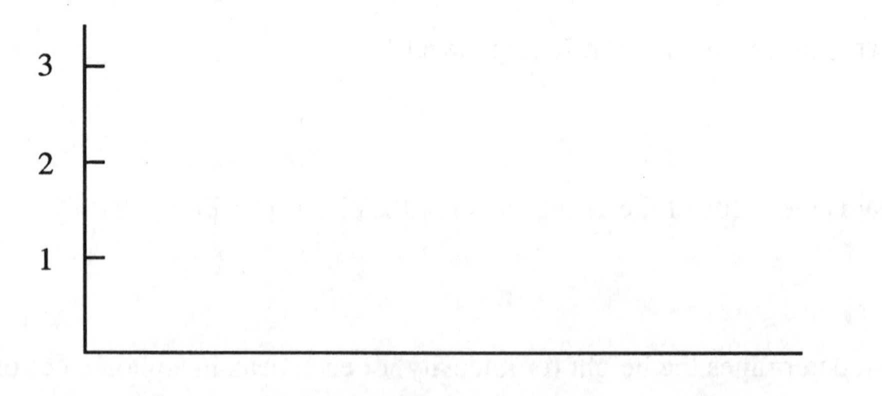

Model 4: A Simulated Photoelectron Spectrum of an "Unknown" Atom.

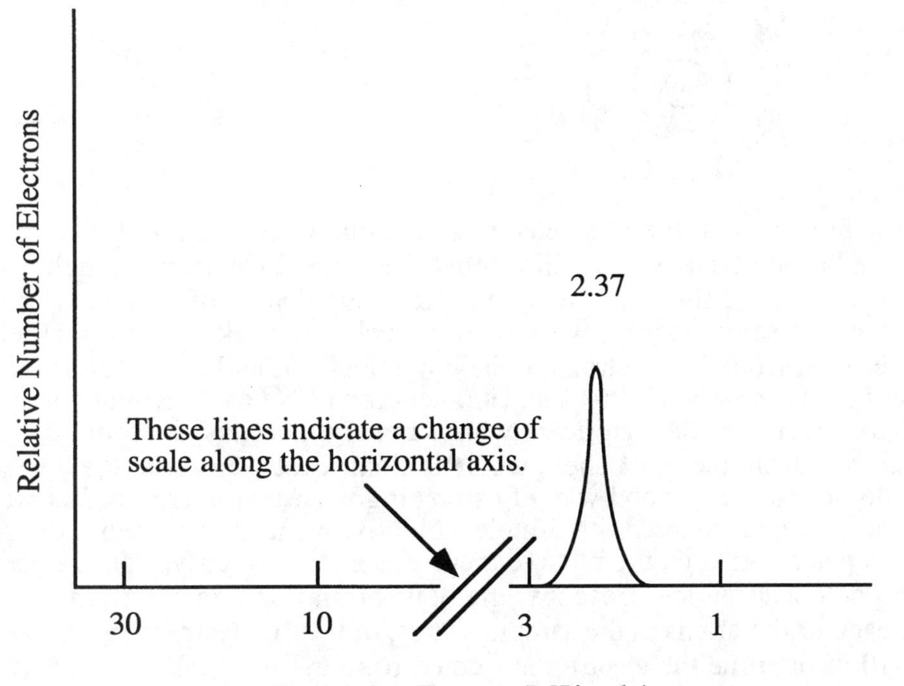

Critical Thinking Questions

12. Based on the number of peaks (one), the intensity of the peak, and your understanding of the shell model:

    a) Explain why it is not possible to determine if the "unknown" atom is H or He?

    b) Explain why the "unknown" atom cannot be Li.

13. Based on the value of the IE given in Model 4 and on the values given in Table 1 of CA 4, identify the "unknown" atom.

Model 5: The Neon Atom.

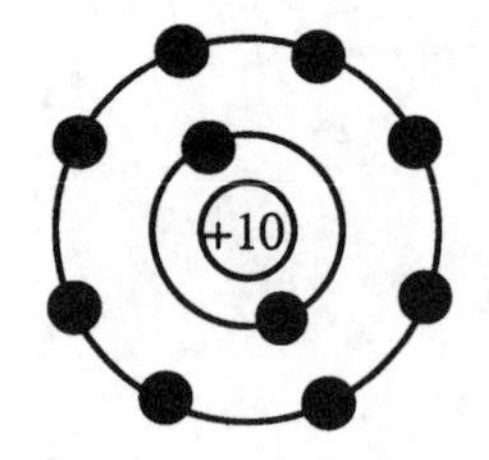

Let us now predict what the photoelectron spectrum of Ne will look like, based on our current model of the Ne atom. In this model, there are 2 electrons in the n =1 shell, and 8 electrons in the n = 2 shell of a Ne atom. Assuming that all of the electrons in each of the shells has the same energy, we would expect two peaks in the photoelectron spectrum. One peak, from the electrons in the n = 2 shell, should appear at an energy of 2.08 MJ/mole, because that is the first ionization energy of Ne as determined previously. The second peak should be at a significantly higher energy, because it corresponds to the ejection of electrons from the n = 1 shell, which is significantly closer to the nucleus. At this point we do not have any good way of estimating what that energy is, but we know that it will be a lot higher than 2.08 MJ/mole. Finally, we also can predict the relative sizes of the two peaks—that is, the relative areas under the two curves on the spectrum. Recall that in photoelectron spectroscopy, the bombarding photon ejects an electron *at random* from each of the atoms in the sample. Thus, of the 10 electrons in Ne, we would expect that 2/10 of the time the electron is ejected from the n = 1 shell, and 8/10 of the time it is ejected from the n = 2 shell. The size of the peak in the spectrum is determined by the relative **number** of electrons with a given IE. Thus, the peak at 2.08 MJ/mole should be 4 times as large as the peak at a much higher energy, which corresponds to the ejection of electrons from the n = 1 shell. *To summarize, our prediction is that the photoelectron spectrum of Ne should consist of two peaks, one at an energy of 2.08 MJ/mole and one at much higher energy, and the relative sizes of these two peaks should be 4:1.*

Critical Thinking Questions

14. The peak due to the n = 1 shell is predicted to be at a much higher ionization energy than the n = 2 peak because the n = 1 shell is "significantly closer to the nucleus." Why is the distance of the shell from the nucleus important in determining the corresponding peak position in the photoelectron spectrum?

15. Why is it expected that 2/10 of the ejected electrons will come from the n = 1 shell, and 8/10 of the electrons from the n = 2 shell?

16. Make a sketch of the predicted photoelectron spectrum of Ne based on the description given above. Indicate the relative intensity (peak size) and positions of the two peaks.

Exercises

1. In a photoelectron spectrum, photons of 165.7 MJ/mole impinge on atoms of a certain element. If the kinetic energy of the ejected electrons is 25.4 MJ/mole, what is the ionization energy of the element?

2. The ionization energy of an electron from the first shell of lithium is 6.26 MJ/mole. The ionization energy of an electron from the second shell of lithium is 0.52 MJ/mole.
   a) Prepare an energy level diagram (similar to the one in Model 3) for lithium; include numerical values for the energy levels.

   b) Sketch the photoelectron spectrum for lithium; include the values of the ionization energies.

3. An atom has the electrons in the energy levels as shown below:

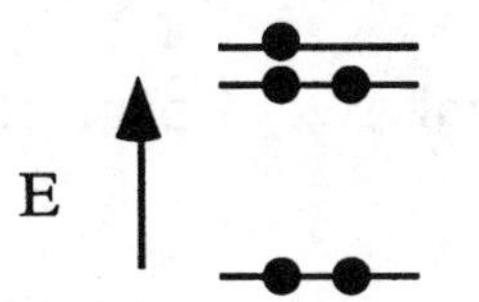

Make a sketch of the PES of this element.

4. J. N. Spencer, G. M. Bodner, and L. H. Rickard, *Chemistry: Structure & Dynamics*, Third Edition, John Wiley & Sons, 2006. Chapter 3: Problem: 78.

ChemActivity 9

The Shell Model (III)

(How Many Peaks Are There in a Photoelectron Spectrum?)

Model 1. Simulated Experimental Photoelectron Spectrum of Neon.

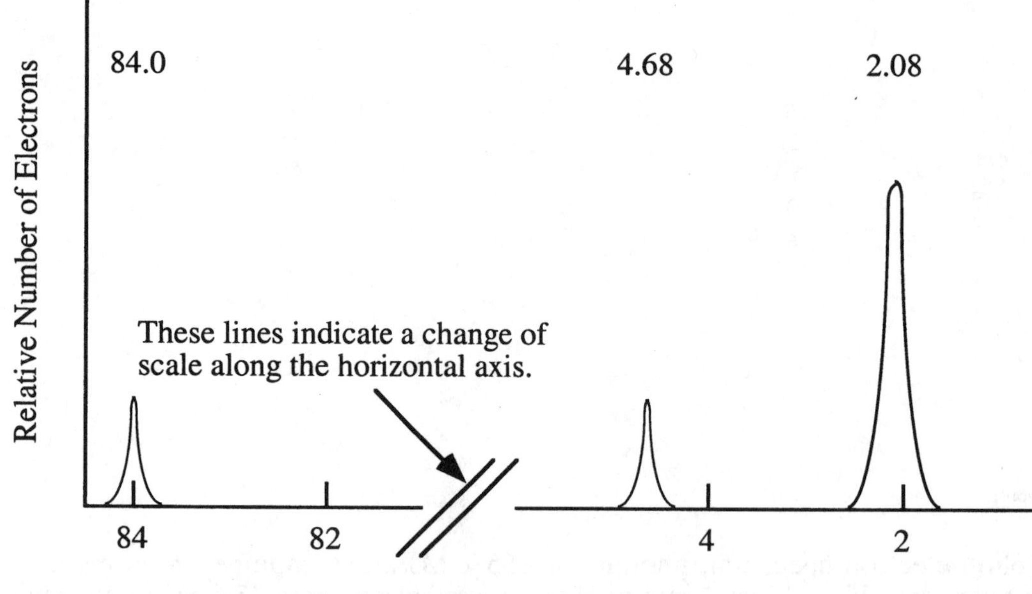

Critical Thinking Questions

1. a) Is the photoelectron spectrum for Ne shown in Model 1 consistent with our model of the structure of atoms?

   b) Describe the ways (if any) in which our prediction was correct and the ways (if any) in which it was not correct.

2. Why is it not important to place numbers along the vertical axis of photoelectron spectra?

3. Based on the spectrum in Model 1, estimate the number of electrons at each of the three energy levels in Ne. Explain your reasoning clearly. (Hint: Recall that the total number of electrons should equal the number of electrons in a Ne atom.)

Information: The Neon Atom Revisited.

Contrary to our predictions, there are *three* peaks in the spectrum, not two! It appears that we must now modify our model to remain consistent with these new experimental results. We note that there is a peak at 2.08 MJ/mole (as expected) and one at a much higher energy—in this case, 84.0 MJ/mole. We can safely assume that this higher energy peak corresponds to the electrons in the $n = 1$ shell, and that the electrons in that shell have an energy of –84.0 MJ/mole. Thus, the other two peaks must both arise from electrons in the $n = 2$ shell. This suggests that there are electrons with two different energies in the $n = 2$ shell, some with an energy of –4.68 MJ/mole, and others with the expected energy of –2.08 MJ/mole. In other words, there are two different energy levels associated with the $n = 2$ shell. To differentiate between them, we label them the 2*s* and 2*p* levels (or subshells), with the *s* designation corresponding to the lower energy level (and higher ionization energy) of the two (IE = 4.68 MJ/mole). (The labels *s* and *p* are used for historic reasons. They do not have any particular significance in the context of our model, but are used to be consistent with the designations used by contemporary practicing scientists.) The lowest energy level of a given shell is always designated as an *s* level, and so the electrons in the $n = 1$ shell are considered to be in a 1*s* energy level. Note that the 2s peak is approximately equal in size to that of the 1s peak, whereas the 2p peak is about three times as big. Thus, we can conclude that there are two electrons (of the eight total) in the 2*s* level of Ne, and six electrons in the 2*p* level.

Our refined shell model of the Ne atom has the 10 electrons distributed in three different energy levels: 2 electrons in a 1*s* level, 2 electrons in a 2*s* level, and 6 electrons in a 2*p* level. At this point we will not be concerned about the details of the differences between the 2*s* and 2*p* levels. The important point is that the 2*s* level is slightly lower in energy than the 2*p*, but not by a large amount. This suggests that the electrons in both levels of the $n = 2$ shell are at nearly the same distance from the nucleus, and are clearly much farther from the nucleus than the electrons in the $n = 1$ shell. Also, we have found that there appears to be a limit of 2 on the number of electrons that can be placed in an s subshell.

Critical Thinking Question

4. What is the reasoning behind the assumption that the peak at 84.0 MJ/mole (for neon; Model 1) corresponds to the electrons in the $n = 1$ shell?

5. Why are two of the three peaks in the spectrum of neon assigned to the $n = 2$ shell, rather than to the $n = 1$ shell?

Model 2. Energy Level Diagrams for Helium, Neon, and Argon.

Helium

1*s* •• −2.37

Neon

2*p* •• •• •• −2.08
2*s* •• −4.68

1*s* •• −84.0

Argon

3*p* •• •• •• −1.52
3*s* •• −2.83

2*p* •• •• •• −24.1

2*s* •• −31.5

1*s* •• −309

Energy levels are not to scale.
Energies are given in MJ/mole.

Critical Thinking Questions

6. Based on the energy level diagram in Model 2, sketch a photoelectron spectrum for Ar. Make sure to indicate the relative intensities and positions of all peaks.

Exercises

1. a) Based on our revised shell model, how many peaks would be expected in a photoelectron spectrum of lithium? b) What would you expect the relative sizes of the peaks to be? c) The *1s* ionization energies for H, He, and Li are 1.31, 2.37, and 6.26 MJ/mole, respectively. Explain this trend. d) The first ionization energies for H and Li are 1.31 and 0.52 MJ/mole, respectively. Explain why the Li first ionization energy is lower.

2. Answer Ex. 1a and 1b for beryllium and for carbon.

3. Sketch the energy level diagram (as in Model 2) for Be and for C.

4. What element do you think would give rise to the photoelectron spectrum shown below? Explain your reasoning.

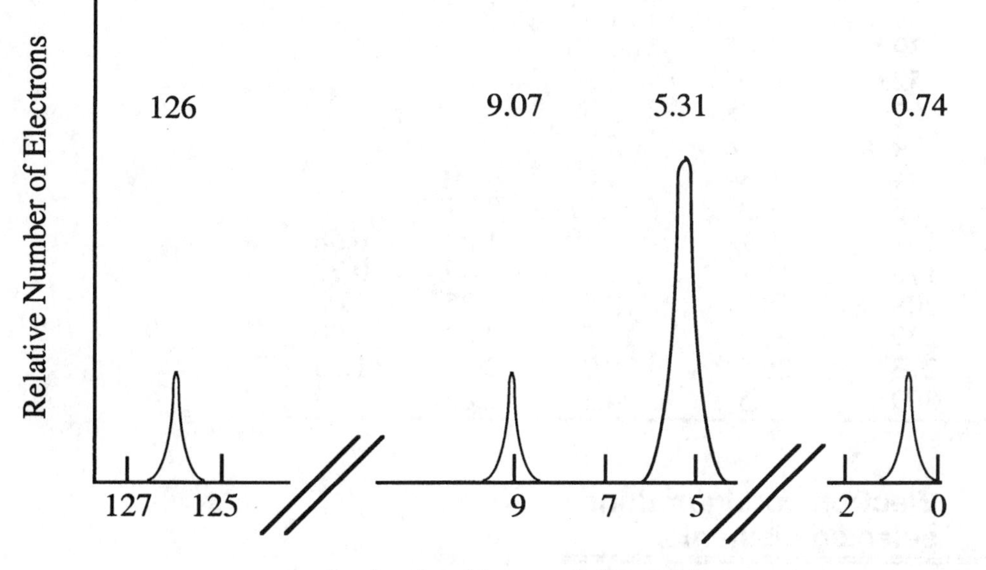

Problems

1. Indicate whether each of the following statements is true or false and explain your reasoning:
   a) The photoelectron spectrum of Mg^{2+} is expected to be identical to the photoelectron spectrum of Ne.
   b) The photoelectron spectrum of ^{35}Cl is identical to the photoelectron spectrum of ^{37}Cl.

2. The energy required to remove a 1*s* electron from F is 67.2 MJ/mole. The energy required to remove a 1*s* electron from Cl is: a) 54 MJ/mole; b) 67.2 MJ/mole; c) 273 MJ/mole; d) a 1*s* electron cannot be removed from Cl. Explain.

ChemActivity 10

Electron Configurations

(How Are Electrons Arranged?)

Model: Ionization Energies and Electron Configurations.

Table 1. Ionization energies (MJ/mole) for the first 18 elements.

Element	1*s*	2*s*	2*p*	3*s*	3*p*
H	1.31				
He	2.37				
Li	6.26	0.52			
Be	11.5	0.90			
B	19.3	1.36	0.80		
C	28.6	1.72	1.09		
N	39.6	2.45	1.40		
O	52.6	3.04	1.31		
F	67.2	3.88	1.68		
Ne	84.0	4.68	2.08		
Na	104	6.84	3.67	0.50	
Mg	126	9.07	5.31	0.74	
Al	151	12.1	7.19	1.09	0.58
Si	178	15.1	10.3	1.46	0.79
P	208	18.7	13.5	1.95	1.06
S	239	22.7	16.5	2.05	1.00
Cl	273	26.8	20.2	2.44	1.25
Ar	309	31.5	24.1	2.82	1.52

Table 2. Electron configurations of selected elements.

Element	Configuration
H	$1s^1$
He	$1s^2$
Be	$1s^2\ 2s^2$
C	$1s^2\ 2s^2 2p^2$
Ne	$1s^2\ 2s^2 2p^6$
Mg	$1s^2\ 2s^2 2p^6\ 3s^2$

Critical Thinking Questions

1. What is the first ionization energy of:

   a) N? b) Ar?

2. Give an experimental method for obtaining the data in Table 1.

3. What information is provided by an electron configuration?

4. What is the relationship between the data in Tables 1 and 2?

5. Is it possible to deduce the electron configuration for an atom from its photoelectron spectrum? If so, describe how. If not, describe why not.

6. We will now construct two possibilities for the photoelectron spectrum of K.

   a) First, consider the first three shells (18 electrons) of K. For these 18 electrons, estimate the IEs [Hint: compare to Ar.] and indicate their relative intensities.

   b) If the 19th electron of K is found in the $n = 4$ shell, would the ionization energy be closest to 0.42, 1.4, or 2.0 MJ/mole? Explain. [Hint: compare to Na and Li.] Show a predicted photoelectron spectrum based on this assumption.

c) If the 19th electron of K is found in the third subshell of n = 3, would the ionization energy be closest to 0.42, 1.4, or 2.0 MJ/mole? Explain. [Hint: compare to other cases in which a new subshell appears.] Show a predicted photoelectron spectrum based on this assumption.

Exercises

1. Explain why more energy is required to remove an electron from the 1s orbital of Na (104 MJ/mole) than to remove an electron from the 1s orbital of Ne (84 MJ/mole).

2. According to the data in Table 1, would it require less than 0.50 MJ/mole, 0.50 MJ/mole, or more than 0.50 MJ/mole to remove a 3s electron from the Mg^{+1} ion? Explain.

3. J. N. Spencer, G. M. Bodner, and L. H. Rickard, *Chemistry: Structure & Dynamics*, Third Edition, John Wiley & Sons, 2006. Chapter 3: Problem: 217.

ChemActivity 11

Electron Configurations and the Periodic Table

(How Many Subshells Are There?)

Model 1. Simulated Photoelectron Spectrum of Potassium.

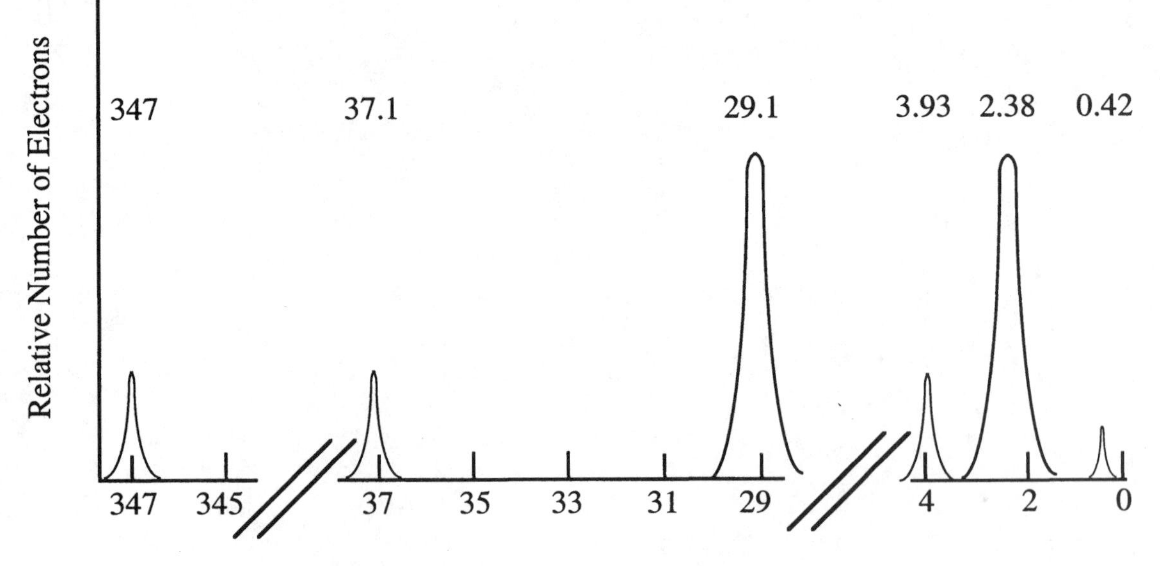

Table 1. Ionization energies (MJ/mole) for selected elements.

Element	1*s*	2*s*	2*p*	3*s*	3*p*	3*d*	4*s*
K	347	37.1	29.1	3.93	2.38		0.42
Ca	390	42.7	34.0	4.65	2.90		0.59
Sc	433	48.5	39.2	5.44	3.24	0.77	0.63

Critical Thinking Questions

1. Which of your predicted spectra from CTQ 6 of ChemActivity 10 provides the better match to the experimental spectrum, Model 1? Explain.

2. Based on the analysis we have used to assign peaks in photoelectron spectra to shells and subshells in atoms, why is the peak at 0.42 MJ/mole in the K spectrum

assigned to the $n = 4$ shell (as opposed to being another subshell of $n = 3$)? Refer to the data in Table 1 of **ChemActivity 10: Electron Configurations.**

Model 2. Simulated Photoelectron Spectrum of Scandium.

(The 1s peak occurs at 433 MJ/mole and is not shown in this spectrum.)

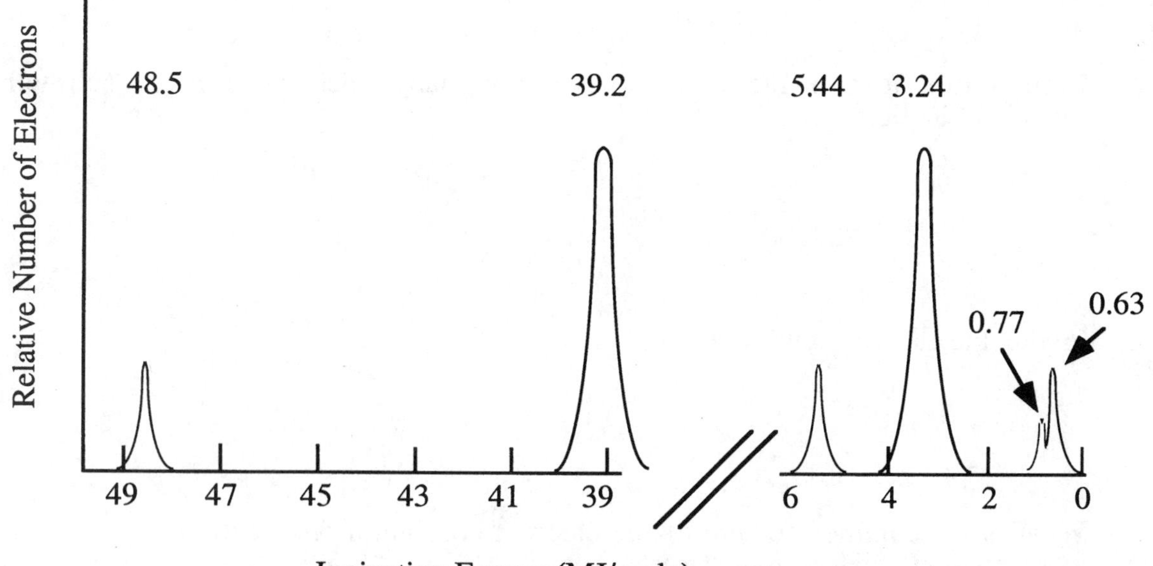

Critical Thinking Question

3. In the photoelectron spectrum of Sc, the peak at 0.63 MJ/mole is assigned to the 4*s* subshell. Why is the peak at 0.77 MJ/mole in the Sc spectrum assigned as a third subshell of $n = 3$ (named 3*d*) as opposed to being a second subshell of $n = 4$ (that is, 4*p*)?

Model 3: The Periodic Table.

Note that the periodic table has an unusual form. The elements are arranged in "blocks" of columns—a block of two columns on the left, six columns on the right, and ten columns in the middle.

Critical Thinking Questions

4. What is the relationship between the form of the periodic table and the electron configurations of the elements?

5. Based on the form of the periodic table, how many electrons is the $3d$ subshell capable of holding?

6. Predict the electron configuration of Ga.

7. What is the common feature of the electron configurations of the elements in a given column of the periodic table?

Exercises

1. Identify the element whose simulated photoelectron spectrum is shown below:

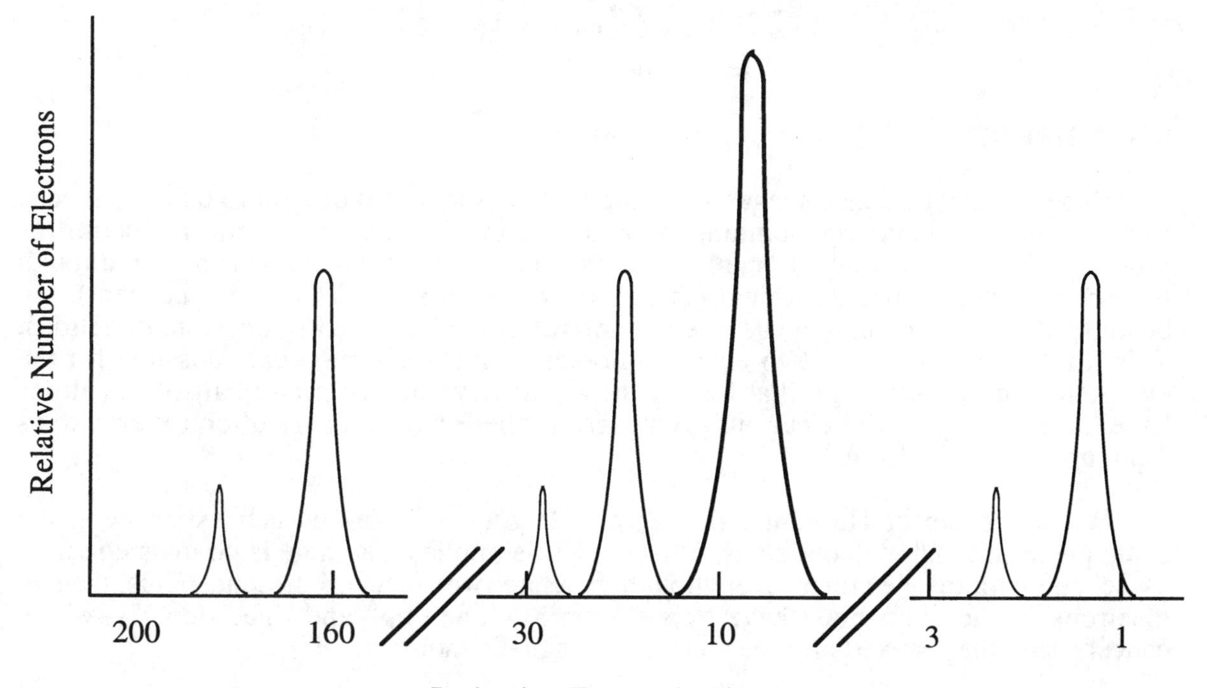

IMPORTANT NOTE: In the above spectrum, the peak which arises from the 1*s* electrons has been omitted.

2. Place the following in order of increasing energy to remove an electron from the 1*s* energy level:

   C Pt Ba Ne Zn Gd

3. Make a rough sketch of the photoelectron spectrum of vanadium. Indicate the subshell that gives rise to each peak and the relative height of each peak.

4. Provide the electron configuration for: P, P^{3-}, Ba, Ba^{2+}, S, S^{2-}, Ni, Zn.

5. How many valence electrons does Ga have?

6. J. N. Spencer, G. M. Bodner, and L. H. Rickard, *Chemistry: Structure & Dynamics*, Third Edition, John Wiley & Sons, 2006. Chapter 3: Problems: 129-131, 134, 137-139, 141.

Problems

1. As atomic orbitals are filled, the 6*p* orbitals are filled immediately after which of the following orbitals? 4*f*, 5*d*, 6*s*, 7*s*.

2. Provide the electron configuration for: Pd, Pd^{2+}, Gd, Gd^{3+}.

ChemActivity 12

Electron Spin

(Are Atoms Magnetic?)

Information

Experimental evidence suggests that the electrons in atoms can act like tiny magnets. Because an electric charge spinning on an axis can produce a magnetic moment, this property of electrons is called "spin." When a beam of hydrogen atoms is passed through an inhomogeneous magnetic field (this is known as a Stern-Gerlach experiment), the beam splits into two components of equal intensity, but deflected in opposite directions. This implies that there are two equal and opposite magnetic moments possible for the electron in the H atom, and that half of the atoms have one type and half of the atoms have the other type. The electrons giving rise to these moments are often referred to as "spin up" and "spin down."

When a beam of He atoms similarly undergoes a Stern-Gerlach experiment, the beam passes through without being deflected. This implies that there is no magnetic field associated with the He atoms, even though there are two electrons present. Thus, the two electrons in the atom must have opposite spins—one "up" and one "down"—which cancel each other out and provide no overall magnetic moment.

Model 1. The Electron Configurations of the Ground States (lowest energy states) of Several Elements.

H	He	Li	Be	B
1*s* ↑	1*s* ↑↓	2*s* ↑	2*s* ↑↓	2*p* ↑ __ __
		1*s* ↑↓	1*s* ↑↓	2*s* ↑↓
				1*s* ↑↓

Critical Thinking Questions

1. What do the arrows in Model 1 represent?

2. What generalization can be made about 2 electrons in a "filled" *s* subshell?

3. For each case, predict the results of a Stern-Gerlach experiment on a beam of atoms. That is, predict whether the atoms will pass through undeflected or will be split into different components.

   a) Li b) Be c) B d) C

Model 2: Diamagnetic and Paramagnetic Atoms.

An atom with an equal number of spin "up" and spin "down" electrons is known as **diamagnetic**, and the atom is repelled by a magnetic field. In this case we say that all of the electrons are "paired." If this is not the case—that is, if there are unpaired electrons—the atom is attracted to a magnetic field, and it is known as **paramagnetic**. The strength of the attraction is an experimentally measurable quantity known as the **magnetic moment.** The magnitude of the magnetic moment (measured in magnetons) is related to (but not proportional to) the number of unpaired electrons present. That is, the larger the number of unpaired electrons, the larger the magnetic moment.

Here are some simulated data concerning this phenomenon:

Table 1. Magnetic moments of several elements.

Element		Magnetic Moment (magnetons)
H	Paramagnetic	1.7
He	Diamagnetic	0
B	Paramagnetic	1.7
C	Paramagnetic	2.8
N	Paramagnetic	3.9
O	Paramagnetic	2.8
Ne	Diamagnetic	0

Critical Thinking Questions

4. Why is the situation of equal numbers of spin up and spin down electrons referred to as all the electrons being "paired"?

5. Based on the data in Table 1, rank the following atoms in terms of the number of unpaired electrons in each atom: B, C, N, O, Ne. Explain your reasoning clearly.

6. a) Make a diagram for C similar to those in Model 2 that shows why C is paramagnetic. Explain how your diagram is consistent with your answer to CTQ 5 and the data in Table 1.

   b) Based on the data provided in Table 1, would you revise your prediction from CTQ 3d of the results of a Stern-Gerlach experiment on a beam of C atoms? If so, in what way? If not, why not?

7. Make a diagram for N similar to those in Model 2 that is consistent with your answer to CTQ 5 and the data in Table 1.

8. Based on the data in Table 1 and your answers to CTQs 6 and 7, do electrons in a given energy level tend to pair or not?

9. Based on the data in Table 1, indicate the number of unpaired electrons in

   a) an O atom? b) a Ne atom?

10. How many "pairs" of electrons are there in a "filled" *p* subshell?

Information

The model that we have developed for the structure of atoms has been further refined. This more sophisticated model, known as the **quantum mechanical model**, retains most of the general features that we have deduced for atomic structure. Within this model, the electrons in atoms occupy specific regions of space known as **orbitals**, with a maximum of two electrons occupying each orbital. There are three orbitals in a *p* subshell and one orbital in each *s* subshell. The idea that the two electrons in a given orbital must have opposite spins was first proposed by Wolfgang Pauli in 1925, and is known as the **Pauli Exclusion Principle.** Most general chemistry texts have some discussion of these ideas. An interesting introduction to the ideas of quantum mechanics can be found in Sections 3.13 and 3.15 of *Chemistry: Structure & Dynamics,* by J. N. Spencer, G. M. Bodner, and L. H. Rickard (Third Edition). You should read the appropriate sections of your text to become familiar with the terms and basic ideas of this model.

Exercises

1. Using grammatically correct English sentences, describe the structure of a ^{13}C atom as completely as you can. Both the nucleus and the electrons should be considered in your description. You may use a diagram (or diagrams) as part of your answer, but you should explain the significance in words.

2. Indicate whether each of the following statements is true or false and explain your reasoning.

   a) The oxide ion, O^{2-}, has the same electron configuration as neon.
   b) In the Si atom, there are no unpaired electrons.
   c) An atom of Si and an atom of S have the same number of unpaired electrons.
   d) In all atoms with an even number of electrons, all of the electrons are paired.

3. The electron configuration of N can be represented as:

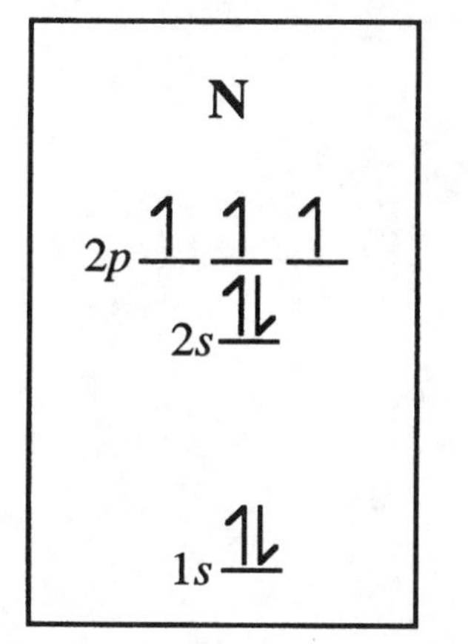

   Why are three separate lines shown for the 2*p* energy level (subshell)?

4. The 3*d* subshell can "hold" ten electrons. Make a diagram, similar to those in the model, for Ni. Predict if the Ni atom will be diamagnetic or paramagnetic.

5. Predict the magnetic moment and the number of unpaired electrons for the F atom.

6. Consider the element X. It has the following properties: X has a smaller atomic radius than Ar; the ion X^- has no unpaired electrons; the ion X^+ has more unpaired electrons than X. What is element X? Explain your reasoning.

7. J. N. Spencer, G. M. Bodner, and L. H. Rickard, *Chemistry: Structure & Dynamics*, Third Edition, John Wiley & Sons, 2006. Chapter 3: Problems: 145, 148, 213 (ignore AVEE).

Problems

1. Which of the following atoms and ions will be paramagnetic? Ti, Ti^{2+}, Na, Na^{+}, Sm, Sm^{3+}, Cl, Cl^{-}.

2. Which species has more unpaired electrons, Fe or Fe^{2+}?

ChemActivity 13

Lewis Structures (I)

(What Makes a Molecule?)

The properties of a molecule depend on how the electrons are distributed in the molecule. For example, it takes more energy to separate an oxygen atom from a carbon atom in a molecule of carbon monoxide, CO, than it does to separate an oxygen atom from a carbon atom in a molecule of carbon dioxide, CO_2. Another example: CO_2 is a linear molecule (the three nuclei lie in a straight line), whereas H_2O is not linear (the three nuclei do not lie in a straight line). These experimentally determined facts can be predicted by making diagrams of molecules, called Lewis structures. The purpose of Lewis structures is to provide a simple way for chemists to represent molecules that allows reasonable predictions to be made about the structure and properties of the actual molecules.

Model 1: Common Methods to Designate Atoms.

Figure 1. Two methods to designate atoms

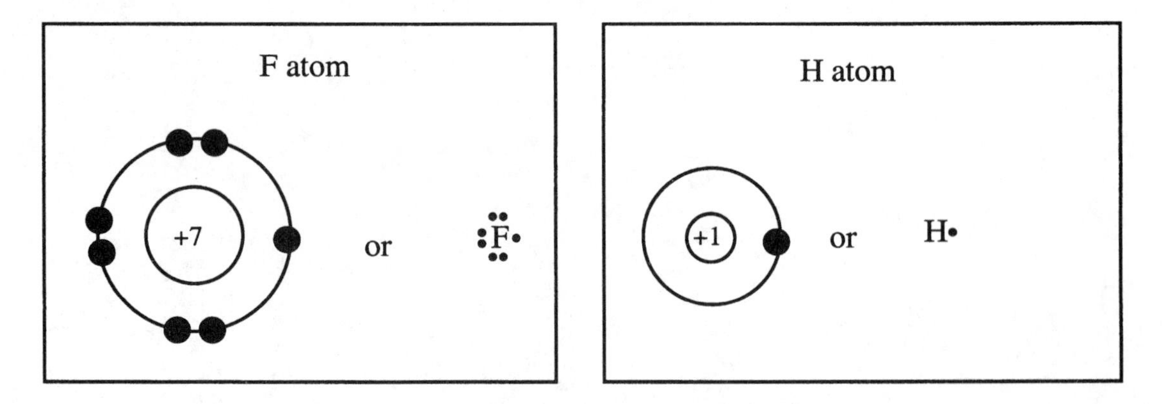

We have seen that a hydrogen atom has a core charge of +1 and that a neutral hydrogen atom has one valence electron. Also, we have seen that a fluorine atom has a core charge of +7 and seven valence electrons. Thus, we have represented these two atoms as shown in Figure 1. Alternatively, we could represent each atom with the appropriate atomic symbol and a dot for each valence electron, also shown in Figure 1. The latter designations take up less space, make the atom easily identifiable, and are more concise; the core charge is not explicit, however, and it is the responsibility of the reader to keep the core charges in mind.

G. N. Lewis proposed the following as representations of the valence electrons for the hydrogen atom and for elements in the groups indicated.

H•

IA	IIA	IIIA	IVA	VA	VIA	VIIA	VIIIA
Li•	•Be•	•Ḃ•	•Ċ• (one dot below)	:Ṅ• (one dot below)	:Ö• (one dot below)	:F̈• (two dots below)	:N̈e: (two dots below)

Critical Thinking Questions

1. For each of the following neutral atoms give the core charge.

   a) Li•

   b) :C̈l• (two dots below)

   c) He:

2. Give the Lewis representation for each of the following atoms.

   a) iodine :Ï· (two dots below)

   b) calcium •Ca•

   c) phosphorus ·P̈ (two dots below)

Model 2: Lewis Structures for Molecules.

The covalent bond—the sharing of two electrons in the valence shell of both atoms:

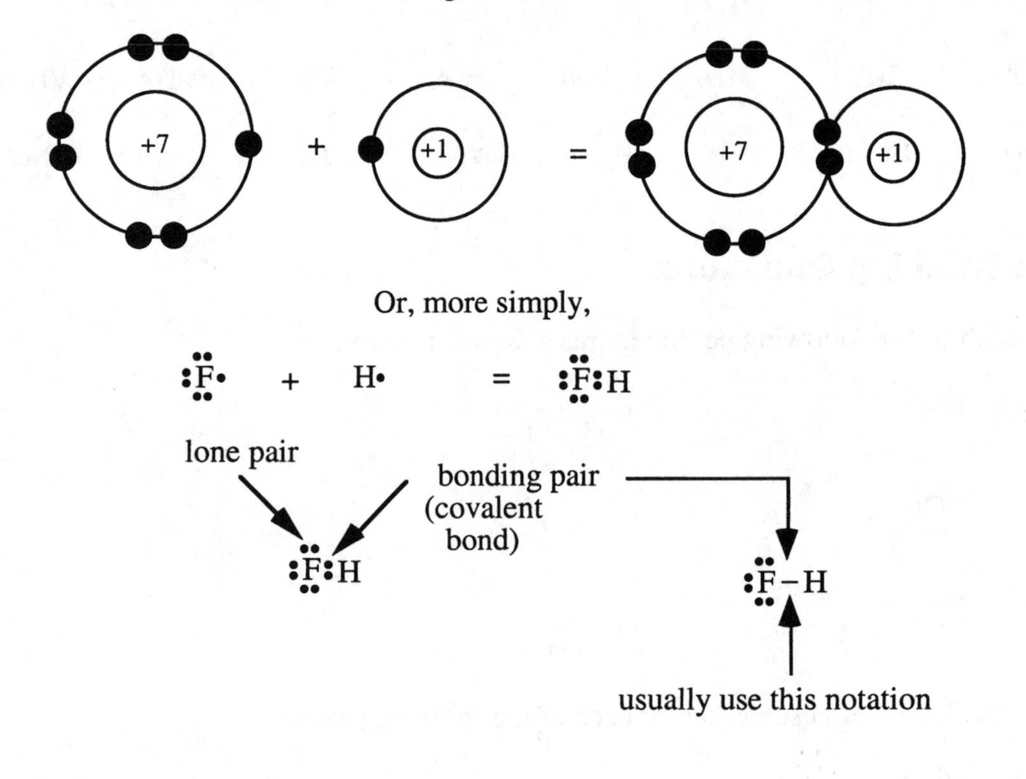

Another example,

:Ö• + 2 H• = :Ö:H or
H

:Ö• + 2 H• = :Ö–H
|
H

Hydrogen must share two electrons—a bonding pair.

The sum of the shared (bonding) electrons and the lone pair electrons for carbon, nitrogen, oxygen, and fluorine atoms must be eight—an **octet**. Usually the other elements in groups IV, V, VI, and VII also follow the octet rule.

Critical Thinking Questions

3. Given the shell model of the atom, suggest a possible reason that Lewis proposed a maximum of two electrons for hydrogen and a maximum of eight for carbon, nitrogen, oxygen, and fluorine atoms?

4. Answer the following for the nitrogen atom:

   a) What is the Lewis representation for N?

   b) How many additional electrons does one N atom require when it forms a molecule?

   c) What is the likely formula for a molecule composed of hydrogen atoms and one nitrogen atom? Draw the Lewis structure for this molecule.

5. What is the likely formula for a molecule composed of hydrogen atoms and one sulfur atom? Draw the Lewis structure for this molecule.

6. Make a checklist that can be used to determine if a Lewis structure for a molecule is correct.

7. Without attempting to draw a Lewis structure, calculate the total number of valence electrons in each of these molecules:

   a) H_2CO

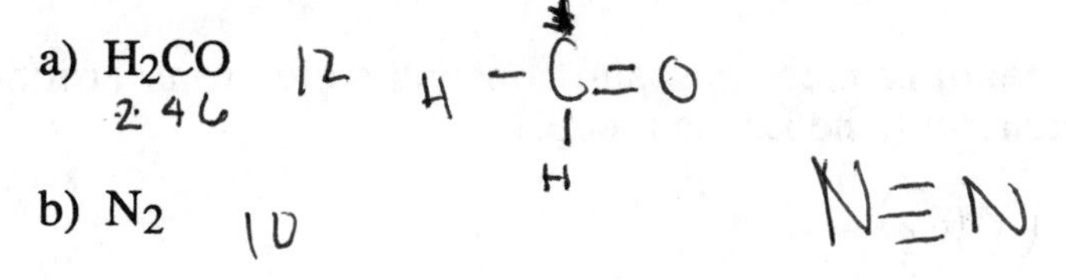

   b) N_2

   c) Cl_2

Model 3: Lewis Structures of Some Molecules.

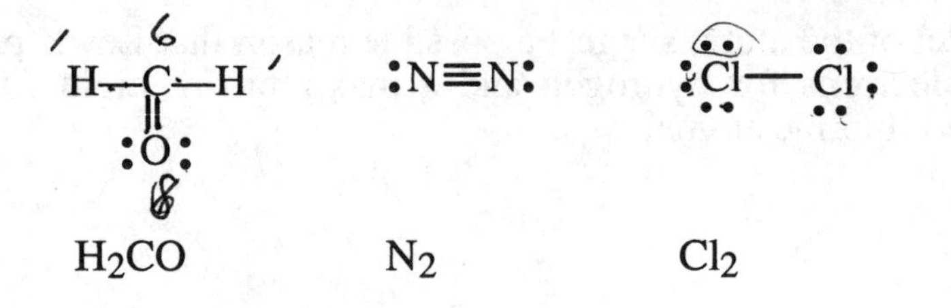

Critical Thinking Questions

8. What is the total number of electrons in the Lewis structure in Model 3 for each molecule:

   a) H_2CO 12

   b) N_2 10

   c) Cl_2 14

9. Compare your answers to CTQs 7 and 8. How does one determine the total number of electrons that should be used to generate a Lewis structure?

   # of valence e⁻

10. For Cl_2, is the sum of the bonding electrons and the lone pair electrons (also known as nonbonding electrons) around each Cl atom consistent with the Lewis model?

    yes

11. For N_2, is the sum of the bonding electrons and nonbonding electrons around each nitrogen atom consistent with the Lewis model?

12. For H_2CO:

    a) Is the sum of the bonding electrons and lone pair electrons around the carbon atom consistent with the Lewis model?

    yes

b) Is the sum of the bonding electrons and lone pair electrons (nonbonding electrons) around the oxygen atom consistent with the Lewis model?

yes

13. Revise (as necessary) your checklist that can be used to determine if a Lewis structure for a molecule is correct.

14. Use your checklist to determine whether or not the following is a correct structure for CO_2:

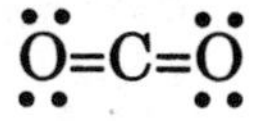

$$\ddot{\underset{\cdot\cdot}{O}}=C=\ddot{\underset{\cdot\cdot}{O}}$$

Correct

Exercises

1. How many valence electrons are in the F_2 molecule? Write the Lewis structure for F_2.
2. How many valence electrons are in the Cl_2 molecule? Write the Lewis structure for Cl_2.
3. How many valence electrons are there in the SiH_4 molecule? Write the Lewis structure for SiH_4.
4. How many valence electrons are there in the $SiCl_4$ molecule? Write the Lewis structure for $SiCl_4$.
5. How many valence electrons are there in the NH_3 molecule? Write the Lewis structure for NH_3. 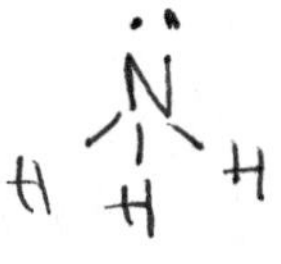
6. How many valence electrons are there in the NCl_3 molecule? Write the Lewis structure for NCl_3.

7. How many valence electrons are there in the PH_3 molecule? Write the Lewis structure for PH_3.

8. How many valence electrons are there in the PCl_3 molecule? Write the Lewis structure for PCl_3.

9. Calculate the total number of electrons in the Lewis structures for each of the following molecules:

   CH_3Cl, CH_3COOH, $HCOOH$, $HNNH$, N_2O_4, $PbCl_4$, C_2H_5OH

10. J. N. Spencer, G. M. Bodner, and L. H. Rickard, *Chemistry: Structure & Dynamics*, Third Edition, John Wiley & Sons, 2006. Chapter 4: Problems: 1, 2, 4, 5, 28abd, 29ab.

Problems

1. How many valence electrons are in the $C_6H_{12}O_6$ molecule?

2. How many valence electrons are in the NH_4^+ ion?

3. How many valence electrons are in the O_2^{2-} ion?

ChemActivity 14

Bond Order and Bond Strength

(Are All Bonds Created Equal?)

Model 1: Single, Double, and Triple Bonds.

Table 1. Bond orders and Lewis structures for selected molecules.

Molecule	Lewis Structure	Bond	Bond Order	Bond Energy (kJ/mole)
H_2	H–H	H–H	1	436
Cl_2	:C̈l – C̈l:	Cl–Cl	1	243
H_2O	:Ö–H with H below O	O–H	1	498
H_3CCH_3	H–C–C–H, each C with H above and below	C–C C–H	1 1	376 420
H_2CCH_2	H–C=C–H, each C with H above	C–C C–H	2 1	720 444
CO_2	Ö=C=Ö	C–O	2	804
H_2CO	H–C–H with C=O below (:O:)	C–O C–H	2 1	782 364
N_2	:N≡N:	N–N	3	945
HCCH	H–C≡C–H	C–C C–H	3 1	962 552

The **bond energy** is defined as the energy required to sever the bond that holds two adjacent atoms together in a molecule. This energy is usually expressed on a molar basis (as the energy to break one mole of specified bonds).

Critical Thinking Questions

1. Verify (using your checklist) that the Lewis structure for H_2CCH_2 given in Table 1 is correct.

2. The title of Model 1 identifies three types of bonds. Give two examples of each type of bond from the molecules in Table 1.

3. What is the relationship between the *bond order* of a bond and the designation of single, double, and triple bonds?

4. What is the relationship between the *bond order* and the number of electrons shared by two adjacent atoms?

5. Rank the three types of bonds of Model 1 in order of increasing strength.

Model 2: Bond Orders and Bond Energies for Selected Molecules.

Molecule	Lewis Structure	Bond	Bond Order	Bond Energy (kJ/mole)
HF	H–F:	H–F	1	570
HCl	H–Cl:	H–Cl	1	432
HBr	H–Br:	H–Br	1	366
HI	H–I:	H–I	1	298
Cl_2	:Cl–Cl:	Cl–Cl	1	243
Br_2	:Br–Br:	Br–Br	1	193
I_2	:I–I:	I–I	1	151
H_3CCH_3	H_3C–CH_3 (H–C–C–H with H above and below each C)	C–C	1	376
H_2CCH_2	H–C=C–H (H above each C)	C–C	2	720
CO_2	O=C=O	C–O	2	804
H_2CO	H–C–H with =O: below C	C–O	2	782
N_2	:N≡N:	N–N	3	945
HCCH	H–C≡C–H	C–C	3	962

Bond length is defined as the distance between the nuclei of two bonding atoms.

Critical Thinking Questions

6. Consider the series HF, HCl, HBr, HI.

   a) What is the bond order for each H–X bond?

   b) What trend is observed in bond energy?

   c) Considering the relative size of F, Cl, Br, and I, what trend would you predict in H–X bond length?

7. Consider the series Cl_2, Br_2, I_2.

   a) What is the bond order for each X–X bond?

   b) What trend is observed in bond energy?

   c) Considering the relative size of Cl, Br, and I, what trend would you predict in X–X bond length?

8. Which of the following statements appears to be true from CTQs 6 and 7? Explain.

   a) The longer the bond, the stronger the bond.

   b) The shorter the bond, the stronger the bond.

9. In Model 2,

   a) What is the range of bond energies for all <u>single</u> bonds?

   b) What is the range of bond energies for all <u>double</u> bonds?

   c) What is the range of bond energies for all <u>triple</u> bonds?

10. Based on your answers to CTQs 6-9 and the data in Model 2, explain how the following conclusion can be reached:

    The most important determinant of bond strength is bond order. If the bond orders are the same, the shorter the bond the stronger the bond.

Exercises

1. Which C–C bond is harder to break?

   $Cl_3C{-}CCl_3$ or $Cl_2C{=}CCl_2$

2. Which C–C bond is harder to break?

   H—C≡C—H or H—C(H)=C(H)—H

3. Which C–N bond is stronger?

   H—C(=N—H)—H or H—C≡N:

4. Using grammatically correct English sentences, describe the relationships between bond order, bond energy, and the number of electrons shared in a bond.

5. The skeletal structures (structures that indicate the arrangement of atoms in a molecule) for formaldehyde and methanol are shown below:

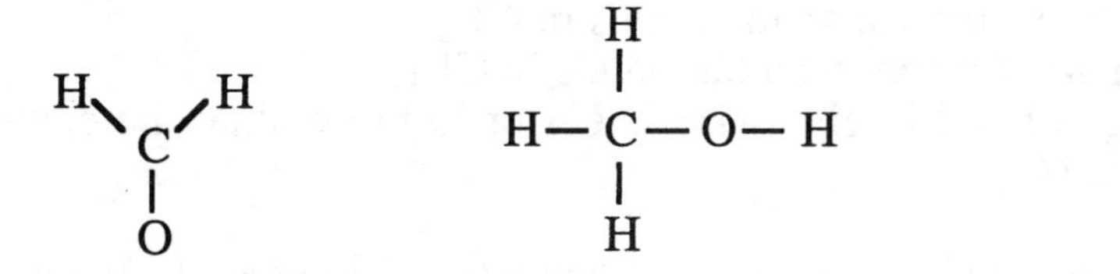

Is the following statement true or false? (Explain your answer.)

It takes more energy to break the C–O bond in formaldehyde than to break the C–O bond in methanol. [Hint: draw the Lewis structure for each molecule.]

6. A rule of thumb states that *about 300 kJ is required to break one mole of single bonds*. Predict the bond energy of double bonds (in kJ/mole) based on this rule of thumb. Predict the bond energy of triple bonds (in kJ/mole) based on this rule of thumb. Compare your predictions to values in Table 1.

7. Consider molecules of the type

```
      H
      |
  H — C — X
      |
      H
```

where X = F, Cl, Br, I.

a) Based on atomic radii, which do you expect to have the longest bond length, C – F, C – Cl, C – Br, C – I? Which do you expect to have the shortest bond length? Explain your reasoning.

b) Which do you expect to have the strongest bond, C – F, C – Cl, C – Br, C – I? Which do you expect to have the weakest bond? Explain your reasoning.

8. For each of the following, which has the stronger bond between the two bold atoms? Give a brief explanation. (Hint: write the Lewis Structure for each molecule.)

a) H_3**CH** and H_3**SiH**
b) $\mathbf{N_2}$ and $\mathbf{O_2}$
c) H_2**PH** and H**OH**
d) H**SH** and H**OH**
e) H**SH** and H**SeH**
f) H_2**PH** and H_2**NH**
g) $\mathbf{F_2}$ and $\mathbf{O_2}$

9. Rank N_2, P_2, As_2, in order from weakest to strongest bond.

10. Indicate whether each of the following statements is true or false and explain your reasoning.

    a) The bonds in NH_3 are shorter than the bonds in NF_3 .
    b) The bonds in CCl_4 are stronger than the bonds in CBr_4 .
    c) The carbon–nitrogen bond in H_3CNH_2 is easier to break than the carbon-nitrogen bond in HCN.

11. J. N. Spencer, G. M. Bodner, and L. H. Rickard, *Chemistry: Structure & Dynamics*, Third Edition, John Wiley & Sons, 2006. Chapter 4: Problems: 47, 49, 120, 126.

Problem

1. Which molecule or ion has the strongest bond between the two atoms: OH^- ; HS^- ; HF ; HCl ; HI ? Explain.

ChemActivity 15

Lewis Structures (II)

(Is One Lewis Structure Enough?)

Model 1: Calculated Bond Orders and Bond Lengths for Selected Molecules.[1]

Molecule		C–C Bond Order (Lewis)	C–C Bond Order (calculated)	C–C Bond Length (calculated) (pm)
ethane,	H_3C-CH_3	1	1.01	150
ethene,	$H_2C=CH_2$ (H–C(H)=C(H)–H)	2	2.00	133
ethyne,	H–C≡C–H	3	2.96	120
benzene,	C_6H_6 (ring of six C, each bonded to one H, alternating single and double C–C bonds)	1 2 1 2 1 2	1.42 1.42 1.42 1.42 1.42 1.42	139 139 139 139 139 139

1 pm = 10^{-12} m

Critical Thinking Questions

1. Why is the bond order (Lewis) for ethene given as "2" in Table 1?
2. How do the calculated bond orders for ethane, ethene, and ethyne compare to bond orders predicted by the Lewis structures?

[1] Bond orders and bond lengths calculated with MOPAC (Oxford Molecular, CAChe). MOPAC calculations yield bond orders and bond lengths that are generally in good agreement with experimental evidence.

3. Based on the data in Model 1 for ethane, ethene, and ethyne, what is the relationship between bond order and bond length?

4. Predict the C–C bond length for a molecule with a C–C bond order of 1.5.

5. How do the calculated bond orders for benzene compare to the bond orders predicted by the Lewis structure?

6. Experimentally, we find that all six C–C bonds have the same bond strength (or bond energy), 509 kJ/mole. Is this fact more consistent with the bond orders predicted by the Lewis structure or with the calculated bond orders?

7. Experimentally (and consistent with the calculated bond lengths in Model 1), we find that all six C–C bond lengths in benzene have the same bond length, 139 pm. Is this fact more consistent with the bond orders predicted by the Lewis structure or with the calculated bond orders? [Hint: refer to your answer for CTQ 4.]

8. What feature(s) of the Lewis structure of benzene, Model 1, is inconsistent with the calculated C–C bond orders in Model 1?

Model 2: Resonance Structures.

An alternative representation for benzene is given in Figure 1.

Figure 1. The resonance hybrid representation of benzene.

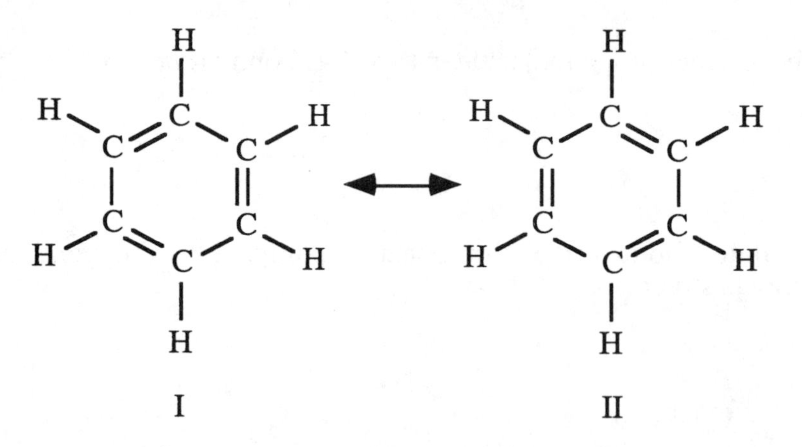

Neither structure I nor structure II alone provides a good description of the true bonding of benzene. Structures I and II are called **resonance structures.** A double-headed arrow is used to indicate resonance structures. Each resonance structure is a legitimate Lewis structure. **The best description of the structure of the molecule is taken to be the <u>average</u> of the resonance structures, sometimes called a resonance hybrid.** For the resonance structures of benzene, above, the C–C bond order is 1.5. That is, the properties of the bond are about halfway between that of a C–C single bond and that of a C–C double bond.

Critical Thinking Questions

9. Based on the resonance hybrid model of Figure 1, why is the C–C bond order (Lewis) 1.5? Explain your analysis.

10. As noted in the footnote for Model 1, the calculated bond orders and bond lengths in Model 1 are expected to be in good agreement with experimental results. Given this assumption, which representation of benzene is better—a single Lewis structure, as shown in Model 1, or the resonance hybrid representation shown of Model 2? Explain your reasoning.

Exercises

1. Which C–C bond is shorter?

```
 :Cl: :Cl:                          :Cl: :Cl:
   |    |                             |    |
:Cl–C—–C–Cl:          or        :Cl– C==C–Cl:
   |    |
 :Cl: :Cl:
```

2. Which C–C bond is shorter?

```
                                    H  H
                                    |  |
H–C≡C–H                 or        H–C=C–H
```

3. Which C–N bond is longer?

```
H–C–H
  ||
 :N
  |
  H                     or        H–C≡N:
```

4. Indicate whether the following statement is true or false and explain your reasoning:

   The carbon-oxygen bond length in H_2CO is the same as the carbon-oxygen bond length in CH_3OH.

5. The molecular formula of cyclobutadiene is C_4H_4 and one of the resonance structures is given below (on the left):

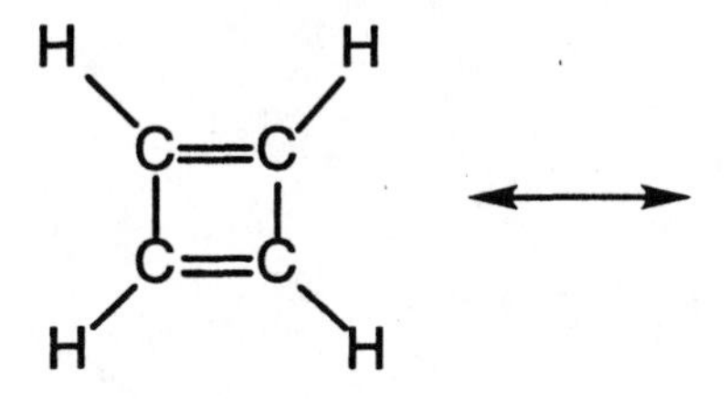

←→

   a) Give the other resonance structure for cyclobutadiene (on the right).

   b) Based on the resonance structure representation, predict the C–C bond order in cyclobutadiene.

   c) Based on the resonance structure representation and the bond length data for ethane and ethene in Model 1, predict the C–C bond length in cyclobutadiene.

Problems

1. Put the following molecules in order of increasing N–N bond length: N_2 ; HNNH ; H_2NNH_2 .

2. Put the following molecules in order of increasing bond length: N_2 ; O_2 ; F_2 .

3. The typical bond length for C–N (bond order = 1) is 147 pm. The typical bond length for C–O (bond order = 1) is expected to be: a) 143 pm b) 147 pm c) 151 pm d) 205 pm e) impossible to estimate from the information given.

4. Is the experimental C–C bond energy for benzene, 509 kJ/mole, consistent with the C–C bond energies given in Model 2 **ChemActivity 14: Bond Order and Bond Strength**? Explain.

ChemActivity 16

Lewis Structures (III)

(Are All Lewis Structures Created Equal?)

Model 1: Two Possible Lewis Structures for CO_2.

$$:\ddot{O}=C=\ddot{O}: \qquad\qquad :\ddot{\underset{..}{O}}-C\equiv O:$$

I II

Experimentally, we find that both C–O bonds in CO_2 are identical. The C–O bond energy in CO_2 is 804 kJ/mole; that is, it requires 804 kJ to break one mole of C–O bonds in CO_2.

Critical Thinking Questions

1. How were the number of electrons in the possible Lewis structures of CO_2 calculated?

2. Is each proposed structure in Model 1 a *legitimate* Lewis structure for CO_2? Explain why or why not.

3. Based on the experimental results, which Lewis structure, I or II, provides a *better* description of CO_2? Explain your reasoning.

Model 2: Formal Charge.

Recall that the purpose of Lewis structures is to provide a simple model from which predictions about molecular structure could be made. Sometimes, as we have seen for CO_2, there is more than one possible Lewis structure for a molecule. The concept of *formal charge* has been found useful for determining the best (most useful) Lewis structure for a molecule. Formal charges are assigned to atoms in molecules according to a set of rules. Specifically,

- formal charge = core charge – number of assigned electrons
- Electrons are assigned as follows:

  1. Nonbonding electrons are assigned to the attached atom.
  2. Shared electrons are evenly divided between the bonded atoms.

- Because of the rules above, the sum of the formal charges on a molecule or ion will equal the total charge on the molecule or ion.

A Lewis structure is not considered complete unless the formal charges are indicated (no "0" is used). The best Lewis structure is the one (or ones if a resonance hybrid) with the lowest formal charges. Formal charges greater than ±1 are never found in good Lewis structures.

Example 1:

For the following Lewis structure of water, H_2O

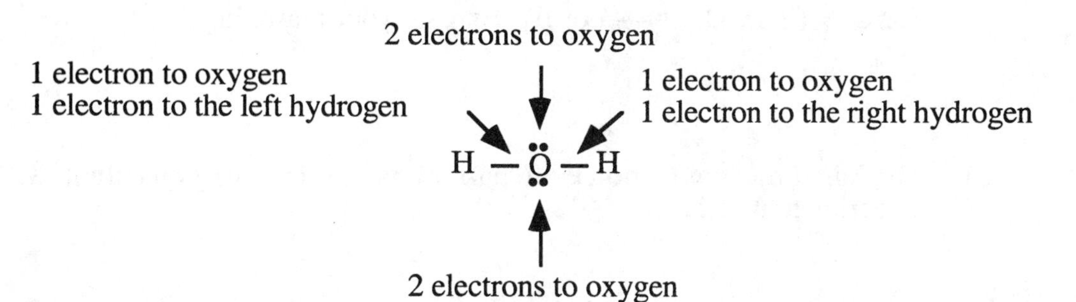

Formal charge on the oxygen = +6 (core charge) – 6 (assigned electrons) = 0
Formal charge on the right hydrogen = +1 (core charge) – 1 (assigned electron) = 0
Formal charge on the left hydrogen = +1 (core charge) – 1 (assigned electron) = 0

All of the formal charges are zero. Zero formal charges are not written on Lewis structures. Thus, the Lewis structure, above, complete with formal charges is:

H— O —H

Example 2:

For the following Lewis structure of carbon monoxide, CO

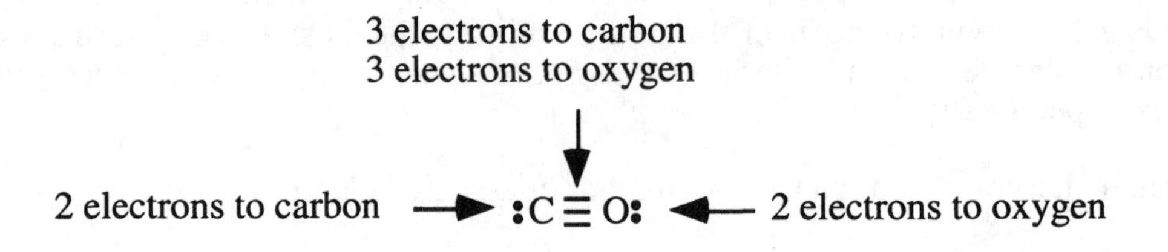

Formal charge on the oxygen = +6 (core charge) – 5 (assigned electrons) = +1
Formal charge on the carbon = +4 (core charge) – 5 (assigned electrons) = –1

The formal charge is circled and written next to the symbol for the atom. Thus, the completed Lewis structure for carbon monoxide, CO, is:

⊖ ⊕
:C≡O:

There is no valid Lewis structure which can be made for CO in which all of the formal charges are zero and the octet rule is obeyed. Thus, the structure shown above is the best Lewis structure for CO because it has the lowest possible formal charges and obeys the octet rule.

Critical Thinking Questions

4. a) Based on the concept of formal charge, which is the better Lewis structure for CO_2 (in Model 1)—I or II? Explain your reasoning.

   b) Explain how your choice in part a) is (or is not) consistent with the experimental data.

5. If the net charge on a molecule is zero, must the formal charge on every atom in the molecule equal zero? Why or why not?

6. Two Lewis structures for formic acid are given below.

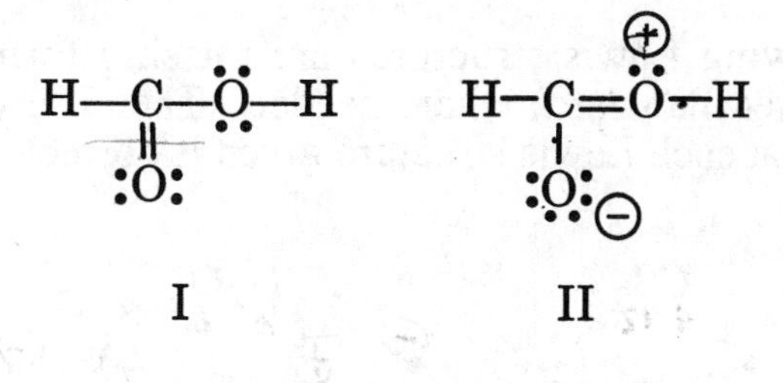

a) According to Model 2, which is the better Lewis structure?

Prefer regular charge

b) The C–O bond length (no hydrogen attached to the oxygen) is 124 pm and the C–O bond length (with a hydrogen attached to the oxygen) is 131 pm. Explain how this is or is not in agreement with your answer to part a.

7. Make a checklist that can used to determine if a Lewis structure is correct and that it is the best Lewis structure.

Exercises

1. Some of the following Lewis structures are missing formal charges. Fill in the formal charges (other than zero) where needed. Then use your list of factors (from CTQ 7) to verify that each Lewis structure given is correct.

NH_4^+ CS_2 N_2O_5 HN_3

2. The following Lewis structure for CO has no formal charges. Explain why this is not a valid Lewis structure.

:C=Ö:

3. The Lewis structure for ozone, O_3, is

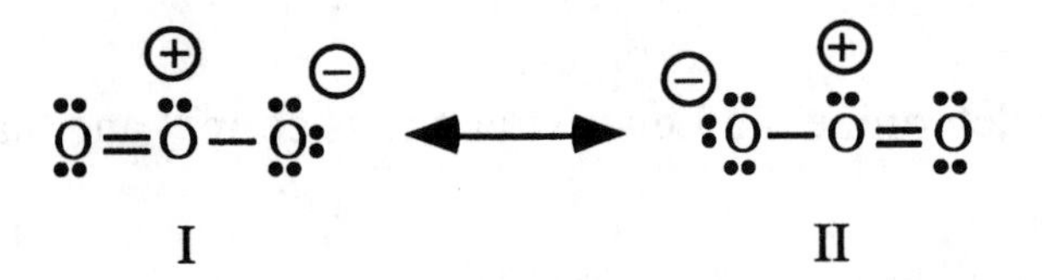

   a) What is the O–O bond order in ozone?
   b) The bond length of a normal oxygen-oxygen single bond is 148 pm. The bond length of a normal oxygen-oxygen double bond is 121 pm. Is the oxygen-oxygen bond length in ozone, 128 pm, consistent with these values? Explain your reasoning.

4. Indicate whether the following statement is true or false and <u>explain your reasoning</u>.

   The carbon-nitrogen bond in H_3CNH_2 is easier to break than the carbon-nitrogen bond in HCN.

5. J. N. Spencer, G. M. Bodner, and L. H. Rickard, *Chemistry: Structure & Dynamics*, Third Edition, John Wiley & Sons, 2006. Chapter 4: Problems: 26, 35, 67abc, 68, 70, 72, 75, 133.

Problem

1. a) Provide the best Lewis structure (include resonance structures and formal charges where necessary) for the acetate anion, CH_3COO^-. Estimate the C–O bond length based on the information given in CTQ 6.

   b) What is the value of the carbon-oxygen bond order in the acetate ion?

ChemActivity 17

Lewis Structures (IV)

(Eight or More Than Eight? That Is the Question.)

Model 1: The Lewis Structure for the Nitrate Ion.

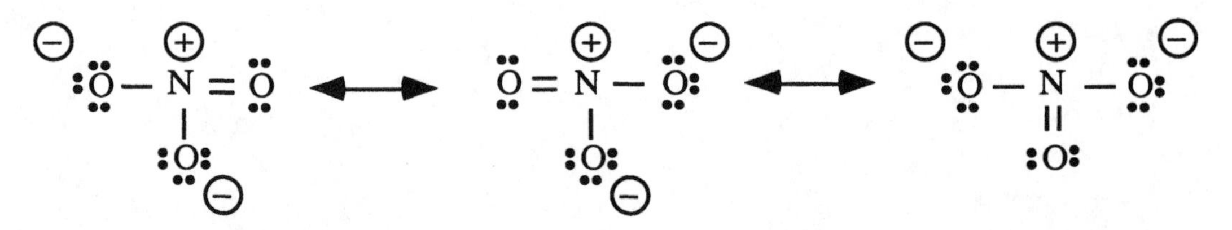

Experimental measurements show that all three N–O bonds in NO_3^- are identical. The N–O bond order is 4/3 for each bond.

Critical Thinking Questions

1. a) How many valence electrons does one nitrogen atom have?

   b) How many valence electrons do three oxygen atoms have?

   c) How many valence electrons does one NO_3 molecule have?

   d) How many valence electrons does one NO_3^- ion have?

   e) How was the number of electrons used for each resonance structure shown in Model 1 calculated?

2. Why is a resonance hybrid representation of NO_3^- better than just a single structure?

3. How was the N–O bond order of 4/3 calculated for NO_3^- ?

Exercises

1. How many valence electrons are in each of the following? SO_3 ; SO_3^{2-} ; SO_4^{2-} ; HSO_4^- ; NH_4^+ ; C_6H_6 ; $C_{14}H_{12}N_4O_2S$; $C_{14}H_{12}N_4O_2S^+$; $C_6H_5NH_3^+$; CCl_4 ; ClO_4^- .

2. How many valence electrons are in the NO_2^- ion? The best Lewis structure for NO_2^- includes two resonance structures. Draw these structures (include formal charges). What is the N–O bond order?

3. Which has the shorter N–O bond length, NO_2^- or NO_3^- ?

4. J. N. Spencer, G. M. Bodner, and L. H. Rickard, *Chemistry: Structure & Dynamics*, Third Edition, John Wiley & Sons, 2006. Chapter 4: Problems: 32, 33, 37, 50-52, 83, 84.

Model 2: Extended Octets.

The sum of the bonding and lone-pair electrons for atoms in the third, fourth, and fifth periods is sometimes greater than eight; see Figure 1.

Figure 1. Atoms in the third, fourth, and fifth periods can have extended octets.

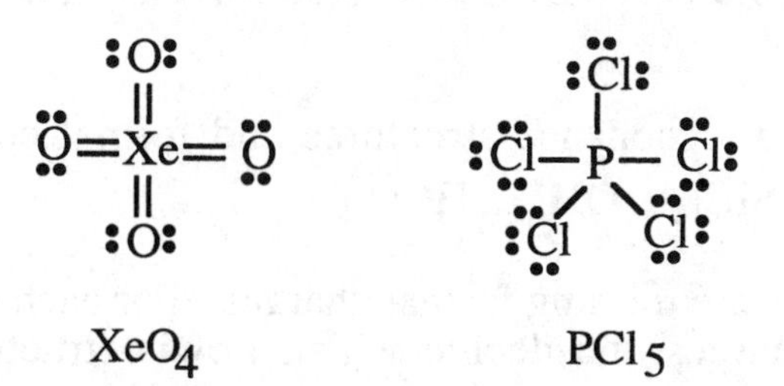

Critical Thinking Questions

4. Confirm that the Lewis structures in Figure 1 are valid and that the formal charges are correct.

5. What is the sum of the bonding electrons and nonbonding electrons for each of the central atoms in Figure 1?

6. Why can C, N, O and F accommodate only eight electrons (sum of the bonding and lone-pair electrons)?

7. Why can atoms in the third (or greater) period accommodate more than eight electrons (sum of the bonding and nonbonding electrons)?

8. Revise (as necessary) your checklist that can be used to determine if a Lewis structure is correct and that it is the best Lewis structure.

Exercises

5. How many valence electrons are in SO_3? Draw the best Lewis structure for SO_3. (Hint: all formal charges are zero in the best Lewis structure for this molecule.) What is the S–O bond order?

6. Draw the best Lewis structure (include resonance structures and formal charges when necessary) for: SO_2; CO_3^{2-} ; $SiCl_4$; OH^- ; HCN.

7. Some of the following Lewis structures are missing formal charges. For each of the following pairs, fill in the formal charges and decide which Lewis structure is better.

SO_4^{2-}

```
      :Ö:                 :O:
       |                   ||
  :Ö — S — Ö:         :Ö — S — Ö:
       |                   ||
      :Ö:                 :O:
```

ClO_4^-

```
      :O:                 :Ö:
       ||                  |
   Ö = Cl = Ö         :Ö — Cl — Ö:
       |                   |
      :Ö:                 :Ö:
```

8. There are six good resonance structures for SO_4^{2-}. One of them is given below. Draw the other five. What is the S–O bond order in SO_4^{2-} ?

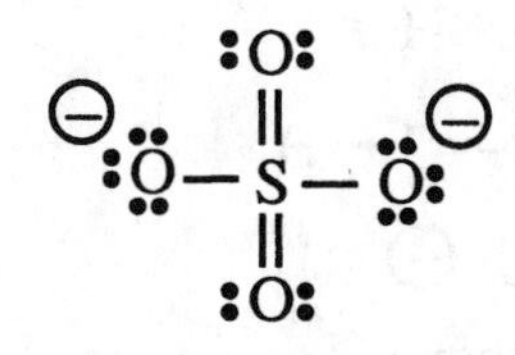

9. Write a Lewis structure for each of the following (include resonance structures and formal charges; skeleton structures are given);

H_2CO (methanal)

```
H— C — H
    |
    O
```

H_2CCH_2 (ethene)

```
   H   H
   |   |
H— C — C — H
```

HCCH (ethyne or acetylene)

```
H— C — C — H
```

NH_2CH_2COOH (glycine, an amino acid)

```
          H
          |
H — N — C — C — O — H
     |    |    |
     H    H    O
```

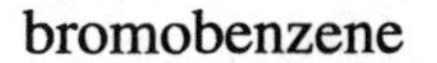

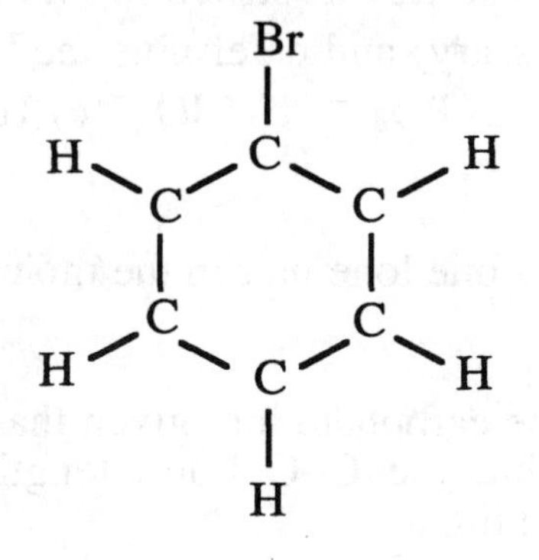

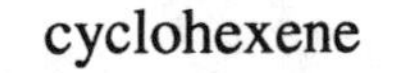

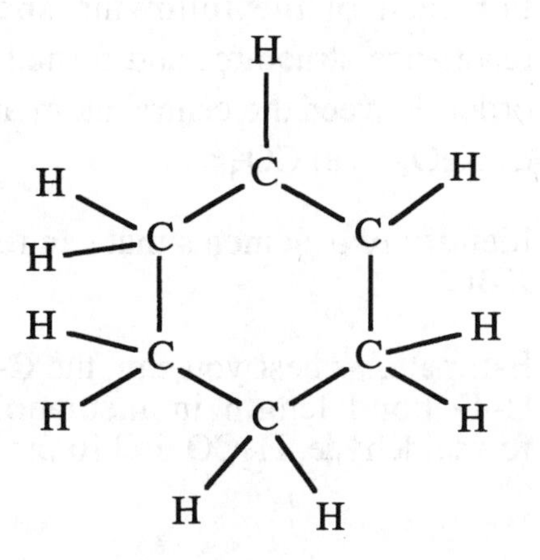

HNO_2 (nitrous acid)

```
H – O – N – O
```

$N_2H_5^+$

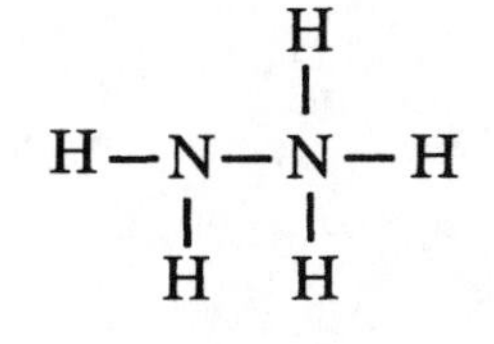

10. Find the error in each of the following Lewis structures. Give the correct Lewis structure.

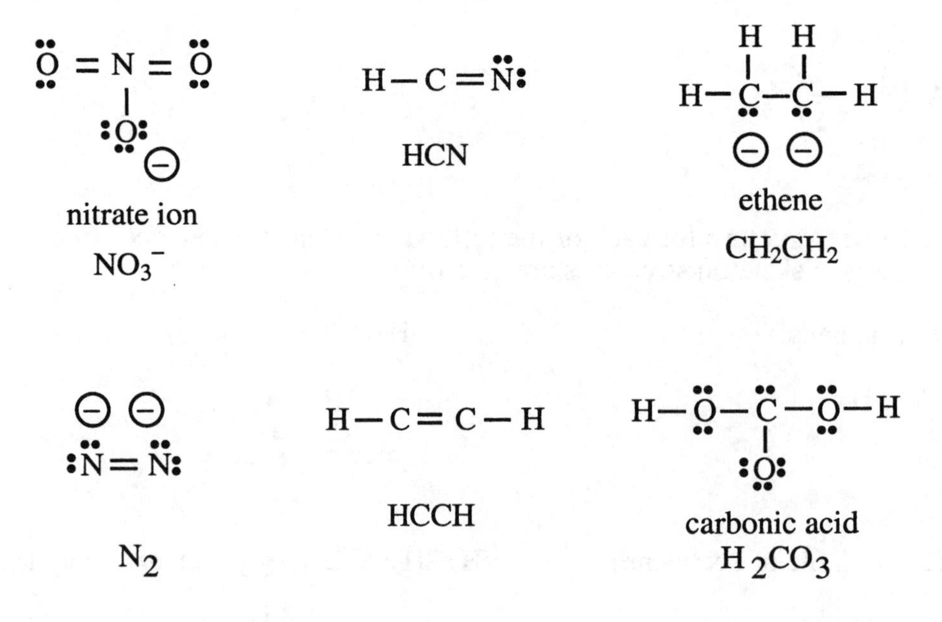

11. J. N. Spencer, G. M. Bodner, and L. H. Rickard, *Chemistry: Structure & Dynamics*, Third Edition, John Wiley & Sons, 2006. Chapter 4: Problems: 117, 122, 132.

Problems

1. For each of the following species, provide the best Lewis structure (include resonance structures and formal charges where necessary) and determine the bond order between the central atom and an attached atom: a) PO_4^{3-} b) ClO_3^- c) $TeCl_4$ d) XeO_4 e) CO_3^{2-}.

2. Identify two elements that can represent "X" if X has one lone pair in the molecule XBr_4.

3. Estimate, as best you can, the C–O bond length in the carbonate ion, given that the C–O bond length in methanol is 143 pm and that the C–O bond length in formaldehyde, H_2CO, is 116 pm. Explain your reasoning.

ChemActivity 18

Molecular Shapes

(What Shapes Do Molecules Have?)

Model 1: Bond Angle and Electron Domains.

A **bond angle** is the angle made by three connected nuclei in a molecule. By convention, the bond angle is considered to be between 0° and 180°.

Table 1. Bond angles and bonding domains in selected molecules.[1]

Molecular Formula	Lewis Structure	Bond Angle (calculated)	No. of Bonding Domains (central atom)	No. of Nonbonding Domains (central atom)
CO_2	O=C=O	∠OCO = 180°	2	0
HCCH	H−C≡C−H	∠HCC = 180°	2	0
H_2CCCH_2	H−C(H)=C=C(H)−H	∠CCC = 180°	2	0
ClNNCl	:Cl−N=N−Cl:	∠ClNN = 117.4°	2	1
NO_3^-	⊖ :O−N(⊕)(=O)−O: ⊖	∠ONO = 120°	3	0
H_2CCH_2	H−C(H)=C(H)−H	∠HCH = 121.1°	3	0

Table 1 continues on the next page

[1]Bond angles calculated with MOPAC (Oxford Molecular, CAChe). MOPAC calculations yield bond orders, bond lengths, and bond angles that are generally in good agreement with experimental evidence.

Molecular Formula	Lewis Structure	Bond Angle (CAChe)	No. of Bonding Domains (central atom)	No. of Nonbonding Domains (central atom)
CH_4	H–C–H with H above and below C	∠HCH = 109.45°	4	0
CH_3F	:F̈: above C; H–C–H; H below C	∠HCH = 109.45° ∠HCF = 109.45°	4	0
CH_3Cl	:C̈l: above C; H–C–H; H below C	∠HCH = 109.45° ∠HCCl = 109.45°	4	0
CCl_4	:C̈l: above C; :C̈l–C–C̈l:; :C̈l: below C	∠ClCCl = 109.45°	4	0
NH_3	H–N̈–H; H below N	∠HNH = 107°	3	1
NH_2F	H–N̈–F̈:; H below N	∠HNH =106.95° ∠HNF = 106.46°	3	1
H_2O	:Ö–H; H below O	∠HOH = 104.5°	2	2

Critical Thinking Questions

1. How is the number of bonding domains on a given atom within a molecule (such as those in Table 1) determined?

2. How is the number of nonbonding domains on a given atom within a molecule (such as those in Table 1) determined?

3. The bond angles in Table 1 can be grouped, roughly, around three values. What are these three values?

4. What correlation can be made between the values in the last two columns in Table 1 and the groupings identified in CTQ 3?

Model 2: Models for Methane, Ammonia, and Water.

Use a molecular modeling set to make the following molecules: CH_4; NH_3; H_2O. (In many modeling kits: carbon is black; oxygen is red; nitrogen is blue; hydrogen is white. Nonbonding electrons are not represented in these models.)

Critical Thinking Questions

5. Sketch a picture of the following molecules based on your models: CH_4; NH_3; H_2O.

6. Describe (with a word or short phrase) the shape of each of these molecules: CH_4; NH_3; H_2O.

Model 3: Types of Electron Domains.

A domain of electrons (two electrons in a **nonbonding domain**, sometimes called a **lone pair**; two electrons in a **single bond domain**; four electrons in a **double bond domain**; six electrons in a **triple bond domain**) tends to repel other domains of electrons. Domains of electrons around a central atom will orient themselves to minimize the electron-electron repulsion between the domains.

Figure 1. Minimization of electron-electron repulsion leads to a unique geometry for two, three, and four domains of electrons.

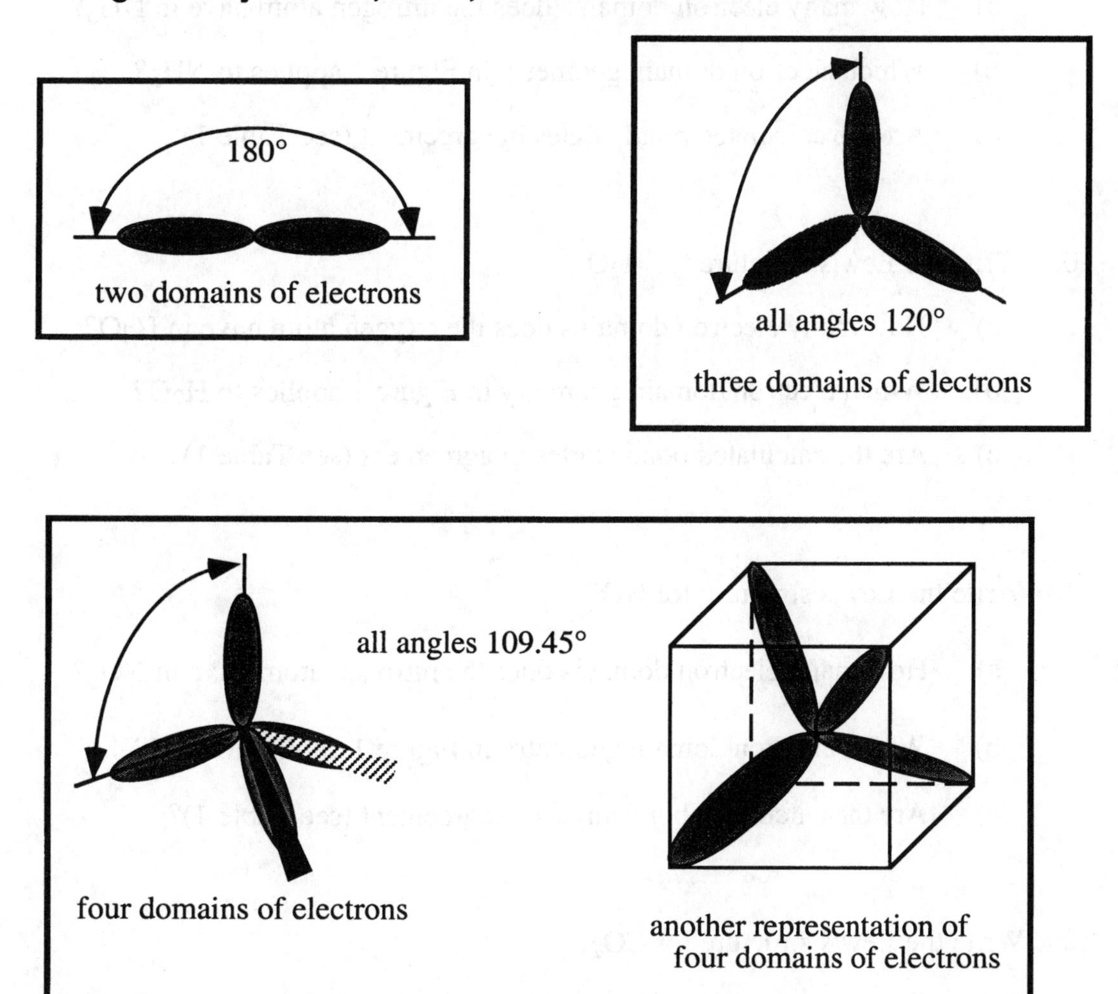

Critical Thinking Questions

7. Based on Figure 1, what bond angle is expected for a molecule containing:

a) two domains of electrons?
b) three domains of electrons?
c) four domains of electrons?

8. Write the Lewis structure for CH_4.

   a) How many domains of electrons does the carbon atom have in CH_4?

   b) Which electron domain geometry in Figure 1 applies to CH_4?

   c) Are the calculated bond angles in agreement (see Table 1)?

9. Write the Lewis structure for NH_3.

   a) How many electron domains does the nitrogen atom have in NH_3?

   b) Which electron domain geometry in Figure 1 applies to NH_3?

   c) Are the calculated bond angles in agreement (see Table 1)?

10. Write the Lewis structure for H_2O.

   a) How many electron domains does the oxygen atom have in H_2O?

   b) Which electron domain geometry in Figure 1 applies to H_2O?

   c) Are the calculated bond angles in agreement (see Table 1)?

11. Write the Lewis structure for NO_3^-.

   a) How many electron domains does the nitrogen atom have in NO_3^-?

   b) Which electron domain geometry in Figure 1 applies to NO_3^-?

   c) Are the calculated bond angles in agreement (see Table 1)?

12. Write the Lewis structure for CO_2.

   a) How many electron domains does the carbon atom have in CO_2?

   b) Which electron domain geometry in Figure 1 applies to CO_2?

   c) Are the calculated bond angles in agreement (see Table 1)?

Information

The names for molecular shapes are based on the position of the atoms in the molecule—not on the position of the electron domains.

Figure 2. The Lewis structure, electron domains, and molecular shape of H_2O.

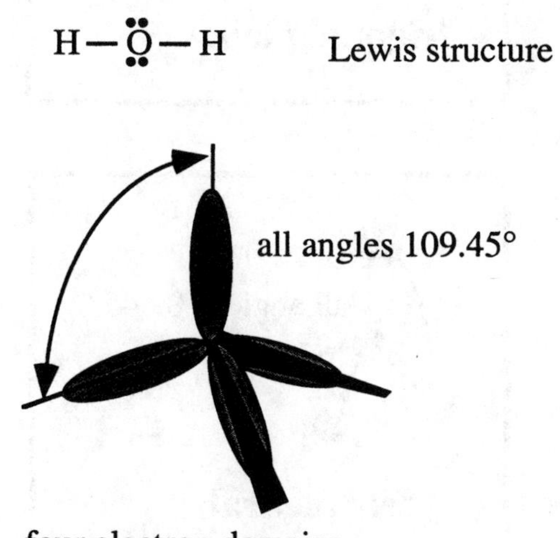

four electron domains

H
O
H

The water molecule is said to be "bent" because the three atoms are not in a straight line. The actual bond angle, determined by experiment, is 105°.

Figure 3. Five common molecular shapes.

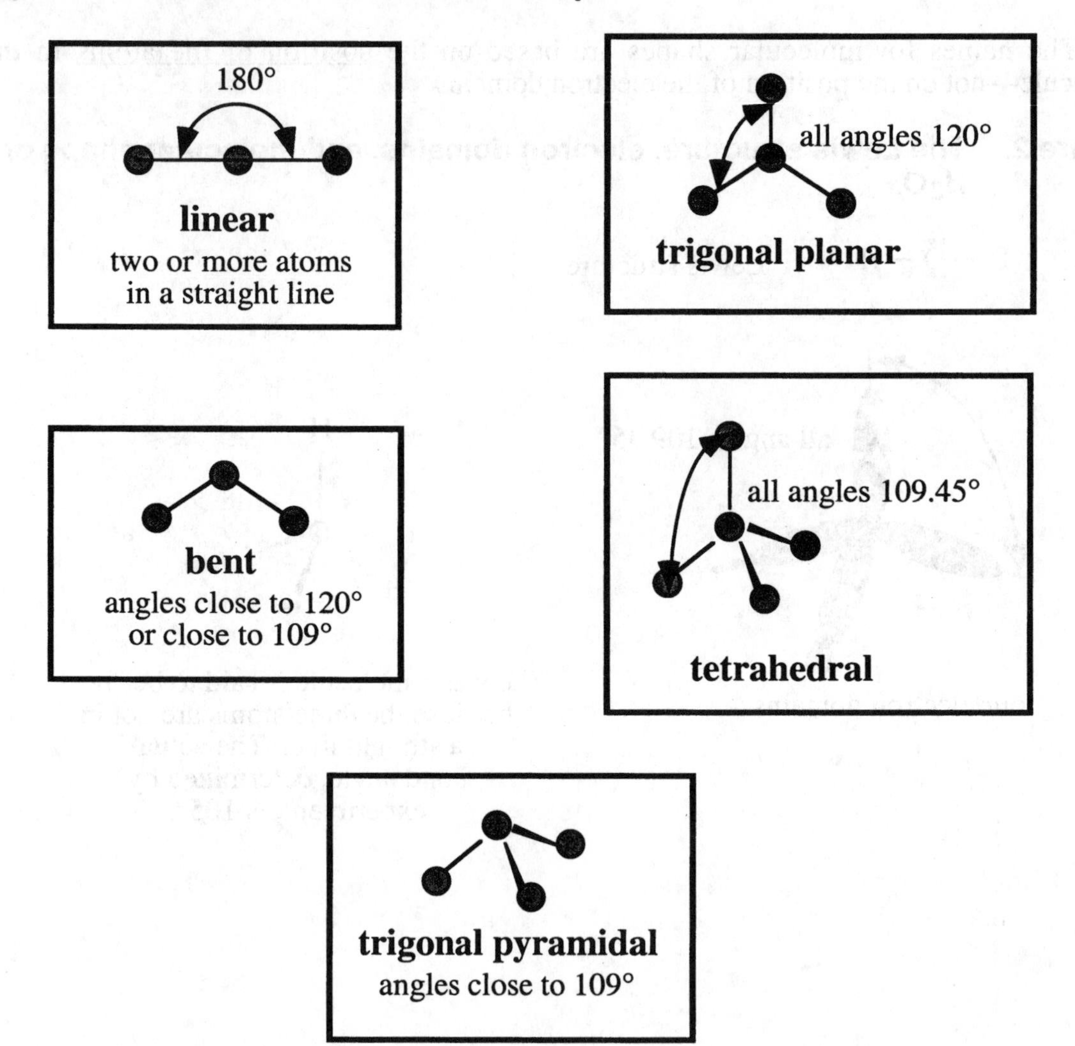

Critical Thinking Questions

13. Considering the geometries described in Figure 1, explain why the bond angle in bent molecules is expected to be close to either 109° or 120°.

14. Using grammatically correct English sentences, explain how the shape of a molecule can be predicted from its Lewis structure.

Exercises

1. Draw the Lewis structure for NH_3. How many electron domains does the nitrogen atom have in NH_3? Make a sketch of the electron domains in NH_3. Examine the drawing for the molecular shape of H_2O given in Figure 2; make a similar drawing for NH_3. Name the shape of the NH_3 molecule and give the approximate bond angles.

2. Draw the Lewis structure for CH_4. How many electron domains does the carbon atom have in CH_4? Make a sketch of the electron domains in CH_4. Examine the drawing for the molecular shape of H_2O given in Figure 2; make a similar drawing for CH_4. Name the shape of the CH_4 molecule and give the approximate bond angles.

3. Draw the Lewis structure for SO_2. How many electron domains does the sulfur atom have in SO_2? Make a sketch of the electron domains in SO_2. Examine the drawing for the molecular shape of H_2O given in Figure 2; note that the electron domain geometry is different for SO_2 and make a similar drawing for SO_2. Name the shape of the SO_2 molecule and give the approximate bond angles.

4. Draw the Lewis structure, sketch the molecules, predict the molecular shape, and give the bond angles for: PH_3 ; CO_2 ; SO_3 ; SO_3^{2-} ; N_2 ; HF ; H_3O^+ ; NH_2F ; CCl_4 ; O_2 ; O_3 ; CO_3^{2-} ; H_2CO.

5. Rationalize $\angle$HCC = 180° in HCCH.

6. Explain why $\angle$ClNN = 117° (close to 120°) in ClNNCl.

7. Predict the bond angles around each atom designated with an arrow in glycine (an amino acid).

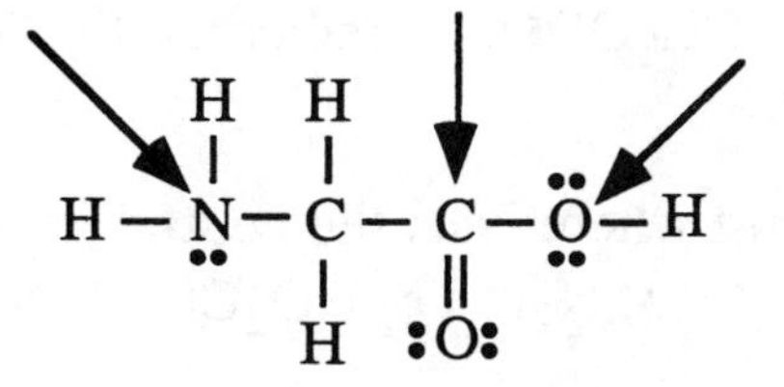

8. Predict the bond angles around each atom designated with an arrow in para-aminobenzoic acid (PABA used in sunscreens).

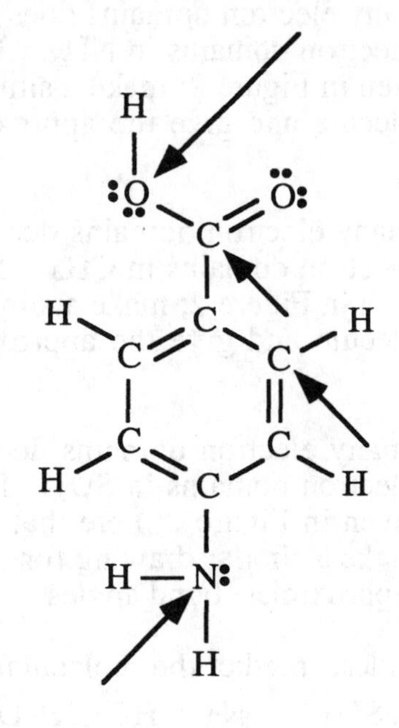

9. J. N. Spencer, G. M. Bodner, and L. H. Rickard, *Chemistry: Structure & Dynamics*, Third Edition, John Wiley & Sons, 2006. Chapter 4: Problems: 87ab, 91 (N_2O is NNO), 96, 102, 130.

Problems

1. Predict the HCC bond angle in the acetate ion, CH_3COO^-. Predict the OCO bond angle in the acetate ion.

2. Predict the bond angle in each of the following where N is the central atom and name the shape of the molecule or ion: a) NO_2^+ b) N_2O c) NO_2Cl d) NH_2^-.

3. Predict the bond angle in H_3O^+. Name the shape of the molecule.

ChemActivity 19

Hybrid Orbitals

Model: Hybridization of Orbitals.

Many chemists describe the arrangement of electrons around an atom in a molecule using different terminology than what we are using. The electron pairs around the central atom in a molecule are said to be in a set of orbitals (domains) that are *hybridized* (constructed) from the usual set of atomic orbitals (*s*, *p*, *d*) for the atom. The construction of hybrid atomic orbitals is beyond the scope of this ChemActivity, but one salient point can be made:

> There is a conservation of orbitals. That is, if one *s*-orbital and one p-orbital are used — two hybrid orbitals (called *sp* hybrid orbitals) are made. The central atom is said to be *sp*-hybridized. If one *s*-orbital and two *p*-orbitals are used — three hybrid orbitals (called sp^2 hybrid orbitals) are made. The central atom is said to be sp^2-hybridized.

Table 1. Characteristics of some hybrid orbitals.

Hybrid Orbitals	Number of Hybrid Orbitals in Set	Component Orbitals	Angle Between Orbitals	Arrangement of Hybrid Orbitals in Set
sp	2	one *s*, one *p*	180°	linear
sp^2	3	one *s*, two *p*	120°	trigonal planar
sp^3	4	one *s*, three *p*	109.45°	tetrahedral

The hybridization assigned to the central atom is based on the experimental bond angle. For example, if the experimental bond angle is about 109° we say that the central atom is sp^3-hybridized. Another example, if the experimental bond angle is about 180° we say that the central atom is *sp*-hybridized.

Critical Thinking Questions

1. The geometry of the methane molecule is shown below. What is the predicted H–C–H bond angle? What is the expected hybridization of the carbon atom?

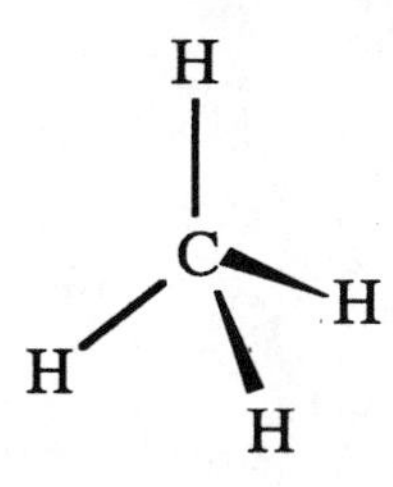

2. The Lewis structure of the water molecule is shown below. What is the predicted H–O–H bond angle? What is the expected hybridization of the oxygen atom?

H — Ö — H

3. One of the resonance structures for the nitrate ion, NO_3^-, is shown below. What is the predicted O–N–O bond angle? What is the expected hybridization of the nitrogen atom?

⊖ ⊕ ⊖
:Ö — N — Ö:
‖
:O:

Exercises

1. For the molecule H_2S:
   a) Draw the Lewis structure.
   b) What is the total number of electron domains about the central S atom?
   c) What is the arrangement of the domains?
   d) What is the predicted bond angle in this molecule?
   e) What is the molecular shape?
   f) What is the expected hybridization of S?

2. Draw the Lewis structure, determine the arrangement of the electron domains, and give the expected hybridization of the central atom in each of the following: H_2S ; CO_3^{2-} ; CCl_4 ; SO_2 ; CO_2 ; PH_3 ; H_2CO.

3. Give the expected hybridization around each atom designated with an arrow in glycine (left; an amino acid) and para-aminobenzoic acid (right; PABA used in sunscreens).

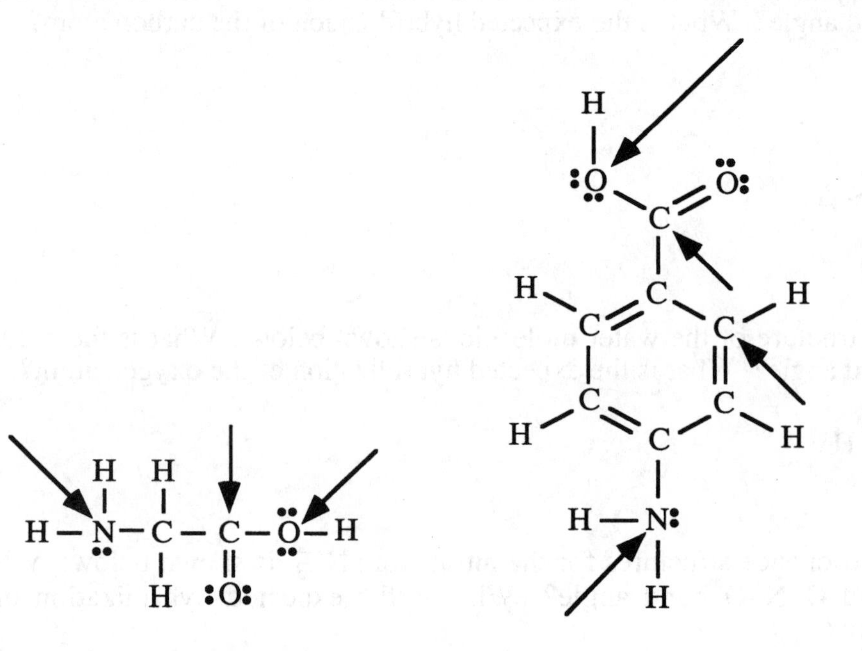

4. J. N. Spencer, G. M. Bodner, and L. H. Rickard, *Chemistry: Structure & Dynamics*, Third Edition, John Wiley & Sons, 2006. Chapter 4A: Problem: 4A-5.

Problems

1. Determine the H–O–O bond angle in hydrogen peroxide, HOOH. Give the hybridization on each oxygen atom.

2. For benzene, give a) the hybridization on all carbon atoms; b) all bond orders; c) all bond angles.

ChemActivity 20

Average Valence Electron Energies

(How Tightly Does an Atom Hold Its Valence Electrons?)

Information

We have noted that the valence electrons play an important role in determining the properties of an element, both physical and chemical. In fact, it is not only the number of valence electrons present in an atom, but also their energies which are important. The ionization energies of the electrons in an atom provide a measure of how tightly the electrons in each shell (and subshell) are held by the atom.

Table 1. Ionization energies (MJ/mole) for the first 21 elements.

Element	1*s*	2*s*	2*p*	3*s*	3*p*	3*d*	4*s*
H	1.31						
He	2.37						
Li	6.26	0.52					
Be	11.5	0.90					
B	19.3	1.36	0.80				
C	28.6	1.72	1.09				
N	39.6	2.45	1.40				
O	52.6	3.04	1.31				
F	67.2	3.88	1.68				
Ne	84.0	4.68	2.08				
Na	104	6.84	3.67	0.50			
Mg	126	9.07	5.31	0.74			
Al	151	12.1	7.19	1.09	0.58		
Si	178	15.1	10.3	1.46	0.79		
P	208	18.7	13.5	1.95	1.06		
S	239	22.7	16.5	2.05	1.00		
Cl	273	26.8	20.2	2.44	1.25		
Ar	309	31.5	24.1	2.82	1.52		
K	347	37.1	29.1	3.93	2.38		0.42
Ca	390	42.7	34.0	4.65	2.90		0.59
Sc	433	48.5	39.2	5.44	3.24	0.77	0.63

Model: Average Valence Electron Energy.

Of particular interest to chemists is a quantitative measure of how tightly an atom holds its valence electrons. We define the *Average Valence Electron Energy* (AVEE) as

$$\text{AVEE} \equiv (aI_s + bI_p)/(a + b)$$

where a and b are the numbers of electrons in the s and p subshells of the valence shell, respectively, and I_s and I_p are the ionization energies of those subshells. For the transition metals (d-block), I_p is replaced by I_d of the preceding shell, and b is the number of d electrons.

Example: Calculate the AVEE for Si and for Sc.

Si: The electron configuration for Si is $1s^22s^22p^63s^23p^2$. The valence shell has two 3*s* electrons (with I_s = 1.46 MJ/mole, from Table 1) and two 3*p* electrons (I_p = 0.79 MJ/mole). Thus,

$$\text{AVEE (Si)} = \frac{(2 \times 1.46 \text{ MJ/mole}) + (2 \times 0.79 \text{ MJ/mole})}{(2+2)} = 1.1 \text{ MJ/mole}$$

Sc: The electron configuration for Sc is $1s^22s^22p^63s^23p^64s^23d^1$. We include the two 4s electrons (with I_s = 0.63 MJ/mole) and one 3d electron (I_d = 0.77 MJ/mole). Thus,

$$\text{AVEE (Sc)} = \frac{(2 \times 0.63 \text{ MJ/mole}) + (1 \times 0.77 \text{ MJ/mole})}{(2+1)} = 0.68 \text{ MJ/mole}$$

In this way it is possible to calculate the AVEE for any element, given the ionization energies of the valence electrons. The calculated AVEEs for several elements are presented in Table 2.

Table 2. Average valence electron energies for several elements (MJ/mole).

H 1.31								He 2.37
Li 0.52	Be 0.90		B 1.17	C 1.41	N 1.82	O 1.89	F 2.30	Ne 2.73
Na 0.50	Mg 0.74		Al 0.92	Si 1.13	P 1.42	S 1.35	Cl 1.59	Ar 1.85
K 0.42	Ca 0.59	Sc 0.68	Ga 1.00	Ge 1.07	As 1.26	Se ≈1.3	Br 1.53	
Rb 0.40	Sr 0.55	Y 0.57	In 0.94	Sn 1.04	Sb 1.13	Te ≈1.2	I 1.35	

Several features of these AVEEs should be noted. First, there is a systematic increase in AVEE as one moves from left to right across each period (with the single minor exception of S). This is consistent with our shell model of the atom, in which the valence electrons are being added to the same outer shell as each period is traversed from left to right, and protons are also added to the nucleus. Thus, on average, the valence electrons in atoms on the right side of the periodic table are held more tightly than those in the same period on the left because, although the radius of the shell is not dramatically altered, the nuclear charge (and core charge) is increased for the elements on the right side of a given period. Second, we note that the elements which tend to form positive ions (the metals) have AVEEs which are relatively small compared to those elements which are identified as nonmetals. For example, the AVEEs for Li and Mg are 0.52 and 0.74 MJ/mole, respectively, while those for O and Cl are 1.89 and 1.59 MJ/mole. As might be expected, those atoms from which it takes relatively little energy to remove valence electrons can most readily form positive ions (cations); if a large amount of energy is needed to remove the electrons in the valence shell (as for Cl and O), then the formation of a negative ion (anion) is more likely to occur.

If we examine those elements which lie along the staircase line in the periodic table separating the metals from the nonmetals (B, Al, Si, Ge, As, Sb, Te) we find that they all have approximately the same AVEE — about 1.16 MJ/mole. In fact, with the single notable exception of Al, all of these atoms have an AVEE within 0.10 MJ/mole of this average value, and no other elements have an AVEE within this range. Thus, it appears that the AVEE of an element can be used to characterize whether that element is a metal or nonmetal. Those elements with AVEEs below 1.06 MJ/mole are identified as metals, and those with AVEEs greater than 1.26 MJ/mole are identified as nonmetals. The elements with AVEEs in the range of 1.06 – 1.26 MJ/mole have properties intermediate between those of metals and nonmetals, and are known as the *semimetals*, or *metalloids*. From this classification, we can see that Al, although it is placed in the periodic table adjacent to the metal/nonmetal dividing line, is properly classified as a metal. This is consistent with the observation that aluminum exhibits all of the characteristic properties of a metal — good conductor of heat and electricity, highly reflective, malleable, etc.

The AVEE of an atom provides a quantitative measure of how tightly valence electrons are attracted to and held on to by an atom. One application of this number is in the quantification of the metallic nature of the elements and an explanation of the placement of the dividing line between metals and nonmetals on the periodic table. There are other ways in which the AVEE of the atoms can be used in understanding and explaining their properties, and the properties of combinations of atoms in compounds. These applications will be discussed later when we examine the structure and properties of molecules.

Critical Thinking Questions

1. a) Why do AVEEs increase as one goes from left to right across a period?

   b) What other periodic property has this same trend?

2. a) Why do AVEEs decrease as one goes down a group?

   b) What other periodic property has this same trend?

3. Why is Be more likely to form Be^{2+} than S is to form S^{2+}?

Exercises

1. Use Table 1 to calculate the AVEE of B and F. Compare your results to the values given in Table 2.

2. Without reference to Table 2, arrange the following in order of increasing AVEE:

   a) P, Mg, Cl

   b) S, O, Se, F

   c) K, P, O

3. Based on position in the periodic table, classify the following elements as metals, nonmetals, or metalloids.

   Ca ; Br ; S ; Si ; Co ; K ; Cu

Problem

1. Estimate the AVEE of P from the following AVEEs: F(2.30); Cl(1.59); O(1.89); S(1.35); N(1.82). Explain your analysis. Compare your estimate to the AVEE of P given in Table 2.

ChemActivity 21

Partial Charge

(Are Electrons Shared Equally?)

Model 1: Partial Charge on an Atom.

We can write the following equation for the calculation of formal charge on atom "a" in the molecule "a–b":

$$\text{formal charge on atom "a"} = V_a - N_a - \frac{1}{2}B_a \text{ ,} \qquad (1)$$

where V_a is the number of valence electrons in atom "a"; N_a is the number of nonbonding electrons on atom "a" in the Lewis structure, and B_a is the number of bonding electrons on atom "a" in the Lewis structure. This calculation assumes that each bonding pair of electrons is shared equally by the two bonding atoms (thus the 1/2 in equation 1). In reality, however, electrons are rarely shared equally. One of the atoms typically has more electron-attracting power, or pull, than the other. As a result, one of the atoms has a residual negative charge, and the other atom has a residual positive charge. It is possible to determine the residual charge (or **partial charge**) experimentally; an estimate of this value can also be calculated by several methods. One simple method is given by equation (2):

$$\text{partial charge on atom "a"} = \delta_a = V_a - N_a - p_a B_a \text{ ,} \qquad (2)$$

where p_a is a measure of the electron pull of atom "a" relative to the pull of the atom to which it is bonded, "b". The value of p_a must be between zero and one.

We have seen that an atom's AVEE is a measure of how tightly an atom holds its valence electrons. Consider the HF molecule. The bonding pair of electrons can be thought of as being in the valence shell of both atoms. We would expect that the fluorine atom in HF would attract the bonding pair of electrons more strongly than the hydrogen atom (AVEE of F = 2.30 and AVEE of H = 1.31). As a quantitative tool, we estimate the electron pull of "a" relative to "b", p_a, as follows:

$$p_a = \frac{\text{AVEE}_a}{\text{AVEE}_a + \text{AVEE}_b} \quad . \qquad (3)$$

For H–F, the electron pull of F relative to H, p_F, is

$$p_F = \frac{2.30}{2.30 + 1.31} = 0.637 \quad . \qquad (4)$$

The electron pull of H relative to F, p_H, is:

$$p_H = \frac{1.31}{2.30 + 1.31} = 0.363 \quad . \qquad (5)$$

Thus, the calculated **partial charges** on the F atom, δ_F , and on the hydrogen atom, δ_H , are:

$$\delta_F = 7 - 6 - 0.637 \times 2 = 7 - 6 - 1.27 = -0.27 \ , \qquad (6)$$

$$\delta_H = 1 - 0 - 0.363 \times 2 = 1 - 0 - 0.73 = +0.27 \ . \qquad (7)$$

Critical Thinking Questions

1. What is the basis of the "7", the "6", and the "2" in equation (6)?

2. When two different atoms share electrons, which atom has the partial negative charge?

3. For HF, why is $\delta_F = -\delta_H$?

4. For any diatomic molecule A–B, what is the following sum: $\delta_A + \delta_B$?

Exercises

1. The AVEE of Br is 1.53 MJ/mole. Use equation (2) to calculate the partial charge on Br in HBr. What is the charge on H in HBr? How do your values compare to the values in Table 1 (below)?

2. The AVEE of Br is 1.53 MJ/mole. Calculate the partial charge on Br in Br_2. Does your answer make sense? Explain.

3. What is the charge on any atom in a homonuclear diatomic molecule?

4. If in the AB molecule $\delta_A = 0.13$, what is δ_B?

Model 2: The Relationship Between AVEE and Partial Charge.

The calculated values for δ_H and δ_F , from equation (2), are consistent with various experimental techniques used to determine how the electrons are distributed in the HF molecule. More sophisticated models, such as MOPAC, yield values as shown in Table 1.

Table 1. Partial charges on atoms in selected molecules (results of MOPAC calculations).

Molecule	Partial Charge	
	H	Halogen
HF	+0.29	–0.29
HCl	+0.17	–0.17
HBr	+0.09	–0.09
HI	–0.01	+0.01

Table 2. Average valence electron energies for several elements (MJ/mole).

Element	AVEE (MJ/mole)
H	1.31
C	1.41
N	1.82
O	1.89
S	1.35
F	2.30
Cl	1.59
Br	1.53
I	1.35

Critical Thinking Questions

5. Note the AVEE values for H and Cl in Table 2. Why is it reasonable that, as shown in Table 1, the hydrogen atom has a positive partial charge and the chlorine atom has a negative partial charge in the HCl molecule?

6. Note the AVEE values for H, F, and Cl in Table 2, and the partial charges in HF and HCl in Table 1. Why is it reasonable that the hydrogen atom has a more positive partial charge in HF than in HCl?

Table 3. Partial charges on atoms within a molecule or ion (results of MOPAC calculations).

Molecule	Partial Charge	
	Central Atom	Each Terminal Atom
methane, CH_4	–0.266	0.066
tetrafluoromethane, CF_4	0.577	–0.144
ammonia, NH_3	–0.396	0.132
trifluoronitrogen, NF_3	0.295	–0.098
water, H_2O	–0.383	0.192
dihydrogen sulfide, H_2S	–0.097	0.048
ammonium ion, NH_4^+	–0.094	0.274
carbonate ion, CO_3^{2-}	0.401	–0.800
nitrate ion, NO_3^-	0.704	–0.568

Critical Thinking Questions

7. Compare the values of the partial charges on the carbon atom in methane and on the carbon atom in tetrafluoromethane. Rationalize the positive and negative aspects of these charges. (Hint: use Table 2.)

8. Compare the values of the partial charges on the nitrogen atom in ammonia and on the nitrogen atom in trifluoronitrogen. Rationalize the positive and negative aspects of these charges.

9. a) Why does the oxygen atom in water have a negative partial charge?

   b) Why does the sulfur atom in dihydrogen sulfide have a negative partial charge?

   c) Why is the oxygen atom in H_2O more negatively charged than the sulfur atom in H_2S?

10. Calculate the sum of the partial charges on all of the atoms in each of the following:

    a) methane

    b) water

    c) ammonium ion

    d) carbonate ion

11. Comment on the relationship (if any) between the charge on a molecule or ion and the sum of the partial charges on all the atoms in the species.

Exercises

5. Write the Lewis structure for CO_3^{2-}. a) Use your Lewis structure to rationalize the negative partial charge on the oxygen atoms. b) Explain why your Lewis structure is consistent with the experimental observation that all three oxygen atoms have the same partial charge.

6. The partial charge on the carbon atom in CI_4 is –0.853. What is the partial charge on each iodine atom?

7. The partial charge on the phosphorus atom in the PO_4^{3-} ion is +2.52. What is the partial charge on each oxygen atom?

8. In the chloromethane molecule, $CHCl_3$, the partial charge on the H atom is +0.16 and the partial charge on each Cl atom is –0.04. What is the partial charge on the C atom?

9. The AVEEs of atoms A, B and C are: A = 3.0; B = 1.0; C = 2.5. Estimate (without calculation) the partial charges on A, B, and C in each of the following molecules. After estimating the partial charges for all of them, calculate the partial charges assuming that neutral atoms A, B, and C have seven valence electrons each.

    a) A–A
    b) C–C
    c) A–B
    d) A–C

10. J. N. Spencer, G. M. Bodner, and L. H. Rickard, *Chemistry: Structure & Dynamics*, Third Edition, John Wiley & Sons, 2006. Chapter 4: Problems: 61, 65, 116.

ChemActivity 22

Polar, Nonpolar, and Ionic Bonds

(Why Is Salt Ionic But Sugar Is Not?)

Model 1: Electronegativities by Linus Pauling.

Linus Pauling examined bonds between homonuclear diatomic molecules (such as H_2 and Cl_2) and bonds in heteronuclear molecules (such as HCl). Bonds between different elements appeared to be stronger. He proposed that the bonding electrons in heteronuclear molecules were not shared equally. That is, he reasoned that in heteronuclear molecules one atom attracted the electrons in the bond more strongly than the other atom. Pauling called the ability of an atom (in a molecule) to attract electrons the **electronegativity** (EN) of the atom. In 1937, he devised a quantitative scale for electronegativity in which fluorine was assigned a value of about 4. This scale has since been refined based on more recent experimental evidence.

Table 1. Electronegativities for selected elements.

H 2.30								He 4.16
Li 0.91	Be 1.58		B 2.05	C 2.54	N 3.07	O 3.61	F 4.19	Ne 4.79
Na 0.87	Mg 1.29		Al 1.61	Si 1.92	P 2.25	S 2.59	Cl 2.87	Ar 3.24
K 0.73	Ca 1.03	Sc 1.2	Ga 1.76	Ge 1.99	As 2.21	Se 2.42	Br 2.69	Kr 2.97
Rb 0.71	Sr 0.96	Y 1.0	In 1.66	Sn 1.82	Sb 1.98	Te 2.16	I 2.36	Xe 2.58

Table 2. AVEE and Electronegativity for Selected Elements.

Atom	AVEE	AVEE × 1.8	EN
H	1.31	2.36	2.30
F	2.30	4.14	4.19
Cl	1.59	2.86	2.87
Br	1.53	2.75	2.69
I	1.35	2.43	2.36

There is, of course, a strong correlation between AVEE and electronegativity. As shown for a few atoms in Table 2, multiplying AVEE values by 1.8 yields numbers very close to electronegativity values. For historical reasons, most chemists use electronegativity values to estimate the electron drawing power of atoms in molecules.

Critical Thinking Questions

1. For HF and HBr, δ_H = 0.29 and 0.09, respectively. Use electronegativities to explain why the partial charge on H in HF is more positive than the partial charge on H in HBr.

2. Describe the trend in EN as one moves from left to right across a period of the periodic table.

3. Describe the trend in EN as one moves down a group of the periodic table.

4. Compare the trends observed in CTQs 2 and 3 to the corresponding trends in ionization energy and explain any observed relationship.

Model 2: Polar, Nonpolar, and Ionic Bonds.

The chemical bond formed between two atoms is generally classified as one of three possible types, as shown in Figure 1.

Figure 1. Three possible bond types.

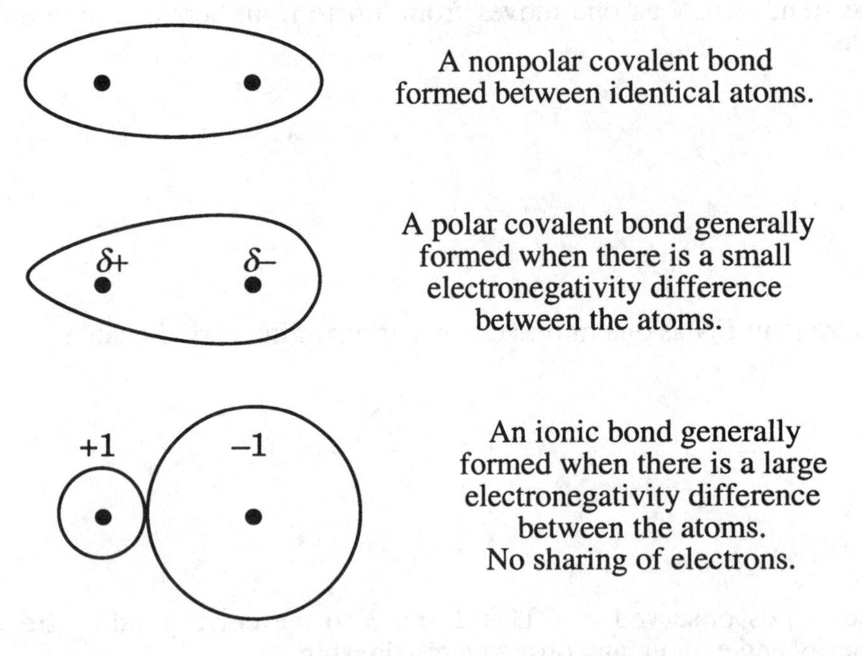

A nonpolar covalent bond formed between identical atoms.
δ+
δ–
A polar covalent bond generally formed when there is a small electronegativity difference between the atoms.
+1
–1
An ionic bond generally formed when there is a large electronegativity difference between the atoms.
No sharing of electrons.

Experimentally, we find that binary compounds usually exhibit ionic behavior (high melting point, 500–3000°C; conduct electricity as liquids) when the electronegativity difference between the constituent atoms is greater than 1.5.

Critical Thinking Questions

5. Why do homonuclear molecules (H_2, Cl_2, N_2, and so on) have nonpolar bonds?

6. When an ionic bond is formed, what type of atom (in terms of relative electronegativity) is likely to:

   a) lose one or more electrons?

   b) gain one or more electrons?

   Explain your reasoning.

Exercises

1. Classify the bond in each of the following molecules (or ions) as nonpolar, polar, or ionic.

   O_2 ; NaF; I_2; KCl ; CO ; NO ; CuO ; CN^-; ICl .

2. Classify each of the following bonds as nonpolar, polar, or ionic.

   C–H in CH_4 ; O–H in H_2O ; Si–Cl in $SiCl_4$; C–H in benzene (C_6H_6)

   Al–O in Al_2O_3 ; Na–F in NaF ; K–Cl in KCl ; N–O in NO_3^- ;

   H–S in H_2S ;

   C–O in

   ```
         H   H
         |   |       ••
     H — N — C — C — O — H
             |   ||  ••
             H  :O:
   ```
   ;

   C–O in $\ddot{\underset{\cdot\cdot}{O}}=C=\ddot{\underset{\cdot\cdot}{O}}$.

3. Excluding the inert gases, which element in the periodic table has the largest electronegativity? Which has the smallest electronegativity?

4. Without referring to a table of electronegativities, identify the most electronegative atom in each case:

   a) Al, P, S, Se, Te
   b) P, Sr, Cu, As, Pb
   c) K, Na, P, As, Si

5. Provide an example of a molecule having polar covalent bonds which are more polar than the bonds in NH_3.

6. J. N. Spencer, G. M. Bodner, and L. H. Rickard, *Chemistry: Structure & Dynamics*, Third Edition, John Wiley & Sons, 2006. Chapter 4: Problems: 56, 57.

Problems

1. Using grammatically correct English sentences, describe what is meant by the term "electronegativity". Give one example of how electronegativity varies in a systematic way in terms of the periodic table, and provide an explanation of that trend in terms of atomic structure.

2. For each of the following, which has the most polar bonds?

   a) CF_4, NF_3, OF_2
   b) OF_2, SF_2, SeF_2

ChemActivity 23

Dipole Moment

(Do Polar Bonds Make Polar Molecules?)

Information

Recall that a polar covalent bond arises when the electronegativity difference between the bonding atoms is sufficient to cause partial charges on the bonding atoms (but not so large as to produce an ionic bond).

Figure 1. Polar and nonpolar covalent bonds.

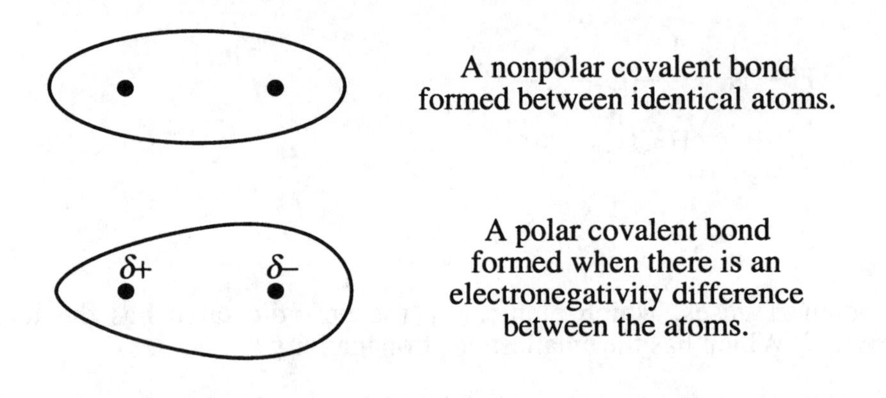

A nonpolar covalent bond
formed between identical atoms.
δ+
δ–
A polar covalent bond
formed when there is an
electronegativity difference
between the atoms.

For example, the electronegativity of H is 2.30 and the electronegativity of Cl is 2.87. MOPAC calculations for HCl yield $\delta_H = +0.17$ and $\delta_{Cl} = -0.17$.

Model: The Dipole Moment.

The polarity of the bond manifests itself in a measurable physical quantity called the **dipole moment**. A dipole moment, **μ**, is a vector quantity that has both magnitude and direction (bold letters are used for vector quantities).

$$\boldsymbol{\mu} = q \times \boldsymbol{d}$$

where q is the magnitude of the two charges (one positive and one negative) and $\boldsymbol{d}$ is the distance (vector) between the two charges. The measured dipole moment for HCl is 1.03 D (D = debye, after the chemist Peter Debye; 1 D = 3.34×10^{-30} C m). The dipole moment can be represented by a vector; a plus sign is used to represent the center of positive charge and the arrow tip represents the center of negative charge. The measured dipole moment for HF is 1.91 D. Bond dipoles and dipole moments for HF, HCl, and H_2O are shown in Figures 2 and 3.

Figure 2. Bond dipoles and dipole moments for HF and HCl.

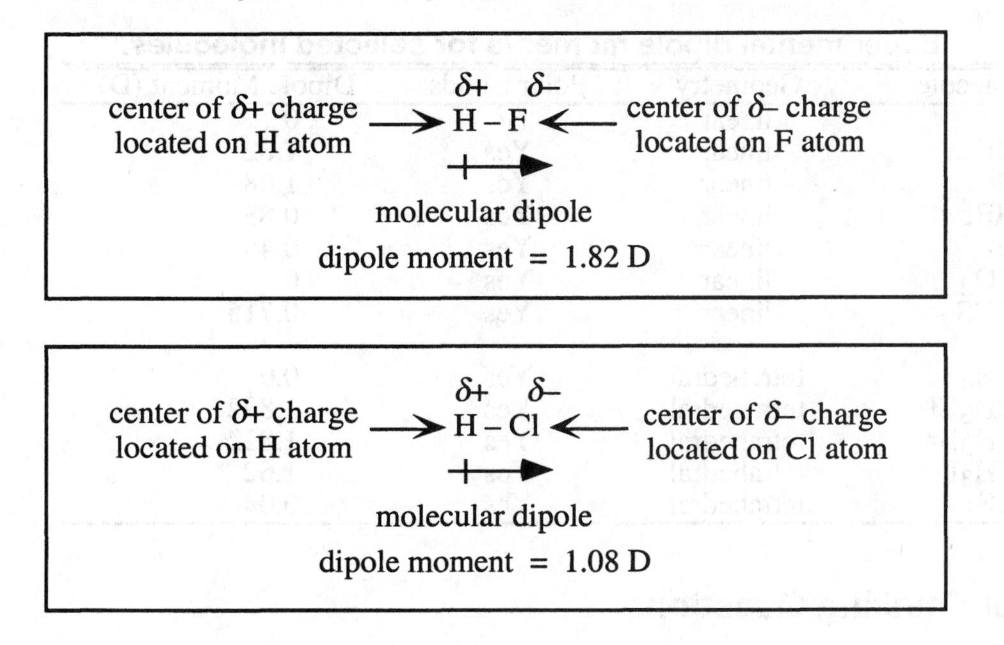

Figure 3. Bond dipoles and dipole moment for H_2O.

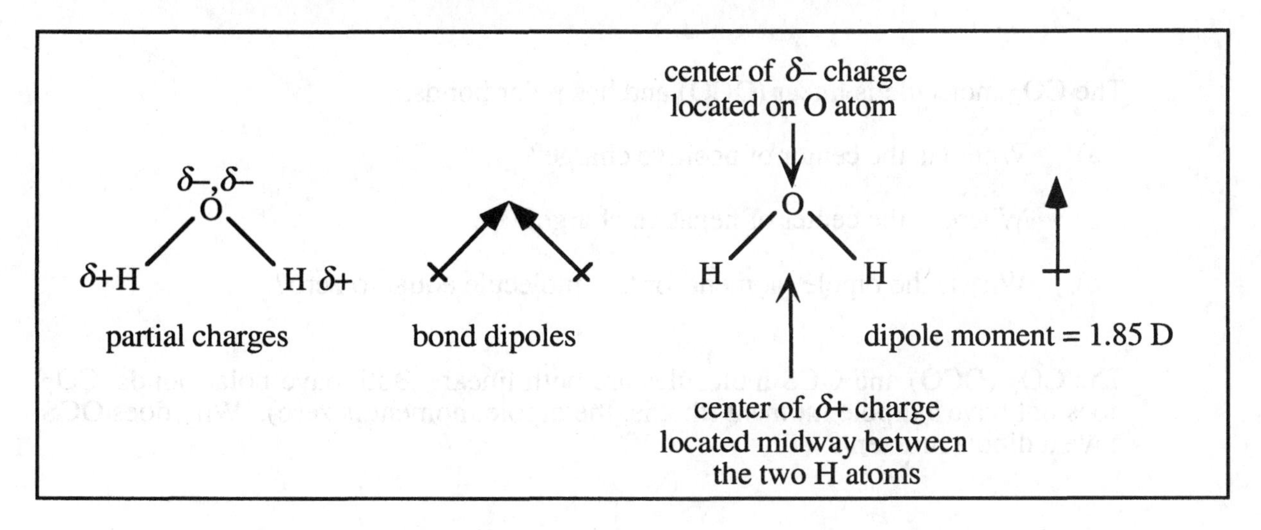

Dipole moments for some other molecules are given in Table 1.

Table 1. Experimental dipole moments for selected molecules.

Molecule	Geometry	Polar Bonds	Dipole Moment (D)
H_2	linear	No	0
HF	linear	Yes	1.82
HCl	linear	Yes	1.08
HBr	linear	Yes	0.83
HI	linear	Yes	0.45
CO_2	linear	Yes	0
OCS	linear	Yes	0.715
CH_4	tetrahedral	Yes	0.0
CH_3Cl	tetrahedral	Yes	1.892
CH_3Br	tetrahedral	Yes	1.822
CH_3I	tetrahedral	Yes	1.62
CF_4	tetrahedral	Yes	0.0

Critical Thinking Questions

1. How was the center of positive charge for the H_2O molecule, Figure 3, determined?

2. The CO_2 molecule is linear (OCO) and has polar bonds.

   a) Where is the center of positive charge?

   b) Where is the center of negative charge?

   c) Why is the dipole moment for this molecule equal to zero?

3. The CO_2 (OCO) and OCS molecules are both linear. Both have polar bonds. CO_2 does not have a dipole moment (that is, the dipole moment is zero). Why does OCS have a dipole moment?

4. Consider the CCl_4 molecule.

   a) Are the C–Cl bonds in CCl_4 polar?

   b) Where is the center of positive charge?

   c) Where is the center of negative charge?

   d) Why is the dipole moment zero for CCl_4?

5. Which has the larger dipole moment in each of the following cases?

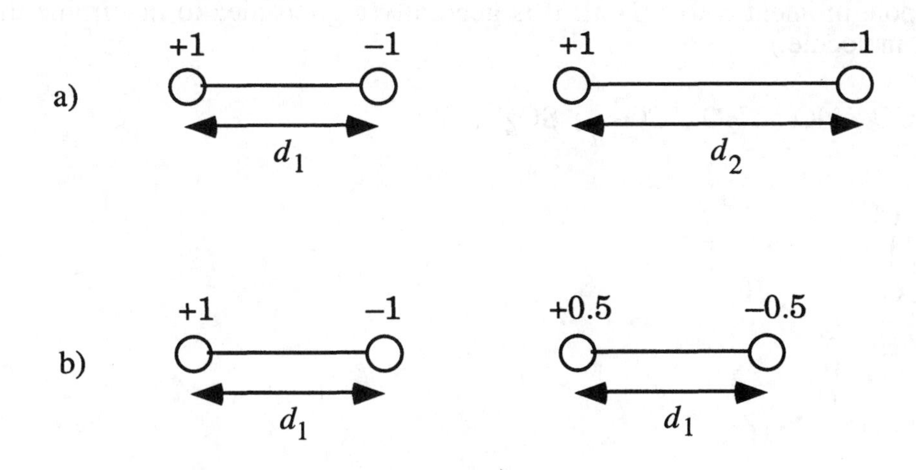

6. Consider HF and HCl:

   a) Which is the bigger atom, F or Cl?

   b) Which has the longer bond length, *d*, HF or HCl?

   c) Which has the greater partial charge—F in HF or Cl in HCl? (Hint: which is more electronegative—F or Cl?)

   d) Why is the dipole moment of HF larger than the dipole moment of HCl? That is, which appears to be the more important factor in determining the dipole moment: the bond length or the partial charge?

7. Based on the dipole moments for HF, HCl, HBr, HI, Table 1, which is more important in the determination of a dipole moment—the bond length (distance) or the electronegativity difference?

Exercises

1. Use the vector notation (see Figure 2 above) to designate the dipole moments of: CO ; HI ; ClF .

2. Write the Lewis structure for NO_3^-. Explain why the dipole moment is zero. (Hint: where does the center of negative charge have to be such that the dipole moment is zero?)

3. Classify each of the following molecules (or ions) as nonpolar (dipole moment = 0) or polar (dipole moment ≠ 0). (Hint: it is generally a good idea to determine the shape of the molecule.)

   O_2 ; I_2; CO ; NO ; H_2O ; CO_3^{2-} ; SO_4^{2-} ; SO_2

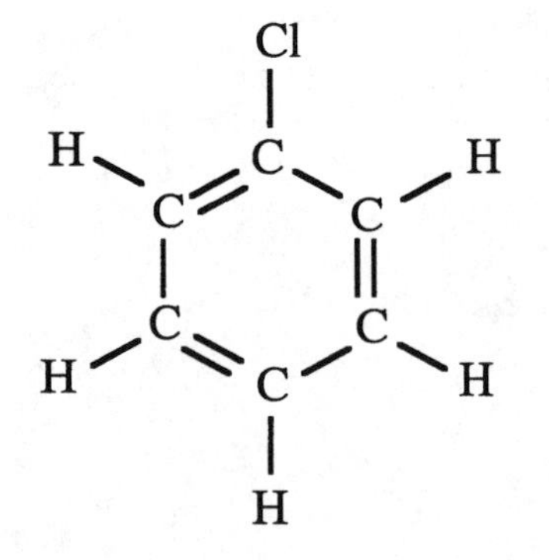

4. Which molecule of each pair has a dipole moment of zero?

   a) N_2 or NO   b) CH_4 or CH_3Cl c) SO_2 or SO_3
   d) NO_3^- or NO_2^- e) SO_4^{2-} or SO_2 f) CO_2 or OCS

5. Indicate whether the following statement is true or false and <u>explain your reasoning</u>.

   The dipole moment of NH_3 is smaller than the dipole moment of CCl_4.

6. Which molecule of each pair has the greater dipole moment?

   a) CCl_4 or CH_3Cl
   b) CH_3Br or CH_3Cl
   c) NF_3 or NI_3
   d) H_2O or H_2S

7. Based on the data in Table 1, predict the dipole moment of CH_3F.

8. For each of the following species: provide the best Lewis structure; sketch the shape of the species including an indication of bond angles; give the hybridization of the central atom; indicate whether or not the molecule is polar (has a nonzero dipole moment).

   a) SCN^- b) PBr_3 c) NO_3^- d) NH_3
   e) CS_2 f) SO_3 g) $CHCl_3$ h) CH_2Cl_2

9. J. N. Spencer, G. M. Bodner, and L. H. Rickard, *Chemistry: Structure & Dynamics*, Third Edition, John Wiley & Sons, 2006. Chapter 4: Problems: 109-112, 128, 129ab.

Problems

1. Which has the larger dipole moment, I or II? Clearly explain.

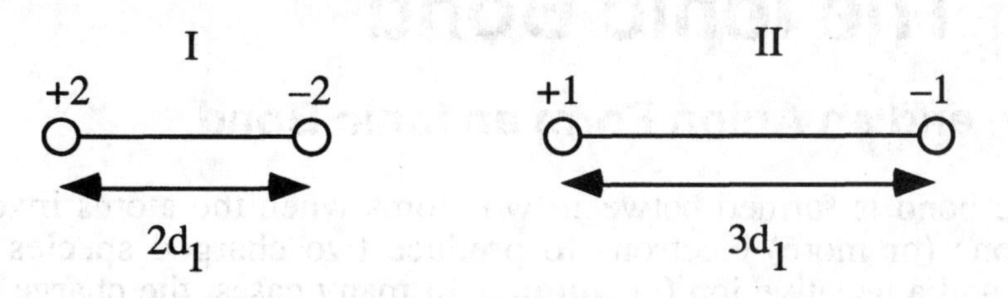

2. For each of the following, which molecule has the largest dipole moment?

   a) CH_3Cl, CH_3Br, CCl_4, CBr_4, CF_4
   b) CO_2, SO_2, NH_4^+, F_2, O_2^{2-}

3. For each molecule (or ion) indicate whether the dipole moment is equal to zero or not: CN^- ; CH_3Br ; H_2O; NO_3^- ; NH_3 .

ChemActivity 24

The Ionic Bond

Model 1: A Cation and an Anion Form an Ionic Bond.

Recall that an ionic bond is formed between two atoms when the atoms involved undergo a transfer of one (or more) electrons to produce two charged species — a positive ion (or **cation**) and a negative ion (or **anion**). In many cases, the charge of an ion can be predicted by considering the electronegativity of the atom and its electron configuration. Atoms which have loosely held electrons tend to form positive ions, whereas atoms which can hold additional electrons relatively strongly tend to form negative ions.

Critical Thinking Questions

1. The ions formed in molecules from Group IA atoms (the alkali metals, such as Li) are almost exclusively M^+ ions rather than M^{2+} ions. Explain this result.

2. The ions formed in molecules from Group IIA atoms (the alkaline earth metals, such as Mg) are almost exclusively M^{2+} ions rather than M^{3+} ions. Explain this result.

3. The ions formed in molecules from Group VIIA atoms (the halogens, such as Cl) are almost exclusively X^- ions rather than X^{2-} ions. Explain this result.

Exercises

1. What is the most prevalent ion of Al? Explain your reasoning.

2. For each of the following atoms, predict the most likely ion:

   S, Cl, Cs, Br, O, Be, N .

3. J. N. Spencer, G. M. Bodner, and L. H. Rickard, *Chemistry: Structure & Dynamics*, Third Edition, John Wiley & Sons, 2006. Chapter 5: Problems: 28, 30.

Model 2: Ionic Bonds and Coulomb's Law.

Recall that in an ionic bond there is no sharing of electrons, but there is a strong attraction between the two ions due to the Coulombic force, as described by equation (1):

$$\text{force} \propto -\frac{q_1 \times q_2}{d^2} \qquad (1)$$

Here, q_1 and q_2 are the charges on the ions and d is the distance between the centers of the two ions. Ionic compounds tend to be solids at room temperature, with high melting points (generally 500 – 3000°C). In contrast with the discrete molecules formed when covalent bonds are present (for example: H_2, H_2O, CH_4), ionic compounds tend to exist as huge, three-dimensional networks of ions of opposite charge that are held together by ionic bonds. Figure 1 below shows an example for the ionic compound consisting of sodium ions and chloride ions. The simplest whole-number ratio of sodium ions to chloride ions is 1:1 in this structure, so the formula for the compound is written as NaCl.

Figure 1. The three-dimensional network structure of NaCl.

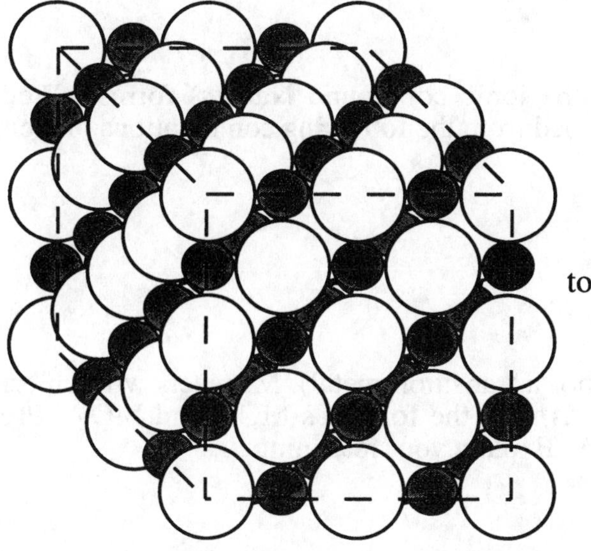

Dashed lines are shown to enhance the three-dimensionality of the structure.

Critical Thinking Questions

4. In NaCl, what are the charges on the ions? Explain your reasoning.

5. In Figure 1, the two different types of ions are represented by spheres of different sizes. Which spheres represent the sodium ions and which spheres represent the chloride ions (the anion formed from the chlorine atom)? (Hint: Consider the electron configurations of the two species.)

6. a) What is the charge on the cation formed from Mg?

   b) What is the charge on the anion formed from O (called the oxide ion)?

   c) What is the formula for the ionic compound containing magnesium ions and oxide ions?

7. a) Which is the larger ion: F^- or Cl^-?

   b) Which would be expected to have stronger ionic bonds: NaCl or NaF? Explain your reasoning. (Hint: see equation 1.)

Exercises

4. When Na(s) and Cl_2(g) react the ionic compound NaCl(s) forms. Predict the formula of ionic compounds formed from the following combinations of elements:

   a) Na(s) and Br_2(g)
   b) Li(s) and O_2(g)
   c) Al(s) and N_2(g)
   d) Mg(s) and Br_2(l)
   e) Ca(s) and O_2(g)

5. A representative group metal (not a transition metal), M, reacts with chlorine and oxygen to form ionic compounds with the formulas MCl_4 and MO_2. Propose a possible identity for the metal M. Explain your reasoning.

Information

One measure of the strength of the bonds holding the ions together in an ionic compound is the melting point: the more strongly the ions are held together, the higher the melting point.

Table 1. Radii, charges, and Coulombic force for some ionic compounds.

Ionic Compound	Radius (pm)		$d = r_{cation} + r_{anion}$ (pm)	Charge		$-\frac{q_1 x q_2}{d^2} \times 10^5$
	cation	anion		cation	anion	
NaF	102	133	235	+1	–1	1.8
NaCl	102	181	283			
MgO	72	140				
MgS	72	184				

Complete Table 1 ("*d*" values, charges, and $-\frac{q_1 x q_2}{d^2} \times 10^5$) before answering the CTQs.

Critical Thinking Questions

8. Consider the ionic compounds NaF and NaCl:

   a) In which compound is the Coulombic force of attraction greater?

   b) NaCl has a melting point of 801 °C. Which of these would you predict is the melting point of NaF: 609 °C, 800 °C, 993 °C? Explain your reasoning.

9. Consider the ionic compounds MgO and MgS:

   a) In which compound is the Coulombic force of attraction greater?

   b) MgO has a melting point of 2852 °C. Which of these would you predict is the melting point of MgS: about 2000 °C, about 2850 °C, about 4000 °C? Explain your reasoning.

Table 2. Melting points of some ionic compounds.

Ionic Compound	Melting Point (°C)
NaF	993
NaCl	801
MgO	2852
MgS	>2000

10. Based on the data in Table 2, which factor, the size of the ions or the charge, has the larger effect on the melting point? Explain.

Exercises

6. Predict which ionic compound has the higher melting point in each of the following pairs:

   a) NaCl and NaBr
   b) NaCl and KCl
   c) MgO and CaO
   d) KCl and CaO
   e) NaCl and MgS
   f) NaCl and $NaNO_3$
   g) KBr and LiF

7. The lattice energy is the amount of energy needed to completely separate (break apart) the ions in one mole of an ionic compound. Indicate which ionic compound is expected to have the larger lattice energy in each of the following pairs:

   a) NaCl and NaBr
   b) NaCl and KCl
   c) MgO and CaO
   d) KCl and CaO
   e) NaCl and MgS
   f) NaCl and $NaNO_3$
   g) KBr and LiF

8. J. N. Spencer, G. M. Bodner, and L. H. Rickard, *Chemistry: Structure & Dynamics*, Third Edition, John Wiley & Sons, 2006. Chapter 5: Problems: 31, 32, 43, 47, 49, 51, 52. Chapter 3: Problem: 211.

Problems

1. Rank the following compounds in order of increasing melting point and explain your reasoning:

   CaO CaS KCl K_2S

2. For each of the following, which compound is expected have the highest melting point?

   a) LiF ; LiCl ; NaF ; NaCl ; KI
   b) NaF ; NaCl ; CaS ; CaO
   c) Na_2SO_4 ; K_2SO_4 ; $CaSO_4$; $BaSO_4$
   d) LiF ; CaO ; BaO; Al_2O_3
   e) H_2O ; NH_3 ; N_2 ; $CaSO_4$; O_2

3. The formula of the ionic compound arising from Al and N_2 is most likely: AlN_2 ; Al_2N_3 ; AlN_3 ; Al_3N ; AlN ?

4. Use grammatically correct English sentences to describe the difference between covalent and ionic bonding.

ChemActivity 25

Metals

(What Makes a Metal Metallic?)

Model 1: Metals, Nonmetals, and Electronegativities.

Some of the properties of **metals** are:

- they have a shine or luster
- they are malleable; that is, they can be hammered or pressed into different shapes without breaking
- they are ductile — they can be drawn into thin sheets or wires without breaking
- they conduct heat and electricity

Nonmetals generally do not have these properties — they are neither malleable nor ductile, and they are often poor conductors of both heat and electricity.

The substantial differences in properties of metals and nonmetals suggest that the structure (that is, the bonding) in the two types of materials is quite different.

Table 1. Classification of several materials composed of single elements listed in order of decreasing electronegativity, EN.

Material	Metal/Nonmetal	State at Room Temperature	EN
$O_2(g)$	Nonmetal	Gas	3.61
$N_2(g)$	Nonmetal	Gas	3.07
$Cl_2(g)$	Nonmetal	Gas	2.87
$H_2(g)$	Nonmetal	Gas	2.30
Cu(s)	Metal	Solid	1.8
Al(s)	Metal	Solid	1.61
Mg(s)	Metal	Solid	1.29
Na(s)	Metal	Solid	0.87

Critical Thinking Questions

1. What type of bonding is present in the nonmetals listed above — covalent or ionic?

2. What type of bonding is present in metals — covalent, ionic, or neither? Explain your reasoning.

3. Based on the data in Table 1, explain how the electronegativity of an element can be used to predict whether the pure substance will be a metal or a nonmetal.

4. Explain, in terms of electronegativities, why the dividing line between metals and nonmetals in the periodic table (the red line that steps down from beneath B to between Po and At) is oriented the way that it is (as opposed to being, for example, horizontal or vertical).

Model 2: The Electronic Structure of Metals.[1]

The bonding in metals is different than both covalent and ionic bonding. In these latter two types of bonding interaction, the electrons in the bond are **localized** — that is, they either are shared by a pair of atoms or they are associated with one of the two species involved in the bonding interaction.

In general, the valence electrons on a metal atom are shared with many neighboring atoms, not just one. In effect, these valence electrons are **delocalized** over a number of metal atoms. Metals tend to exist as extended arrays of spherical atoms that pack so that each atom can touch as many neighboring atoms as possible.

Because the electrons in a metal are not tightly bonded to individual atoms, they are free to move through the metal. A useful picture of the structure of metals envisions the metal atoms as positive ions locked in a crystal lattice surrounded by a sea of valence electrons that move among the ions. The force of attraction between the positive metal ions and the sea of mobile negative electrons forms a **metallic bond** that holds these particles together.

Figure 1. One or more of the valence electrons (per atom) are free to move throughout the metal.

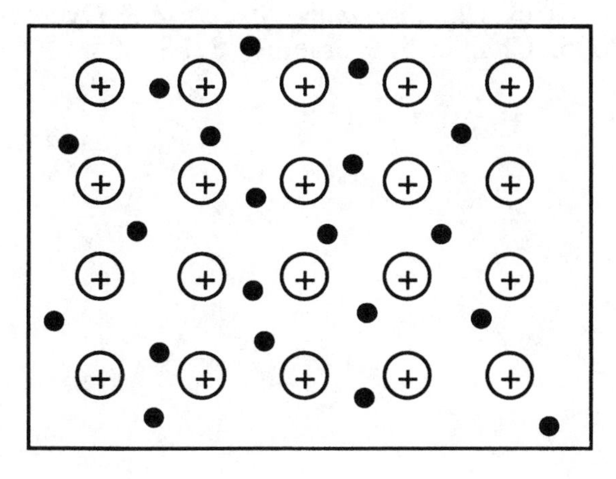

In a metal, the electrons are <u>not</u> bonded to individual atoms; the electrons move rather freely throughout the entire metal. The metal cations form a periodic array of positively charged ions immersed in a "sea" of electrons.

---

[1] Adapted from J. N. Spencer, G. M. Bodner, and L. H. Rickard, *Chemistry: Structure & Dynamics*, Third Edition, John Wiley & Sons, 2006, Chapter 5, Section 10.

Critical Thinking Questions

5. Is the description of metals as being comprised of atoms with loosely held electrons consistent with the data in Table 1? Explain your reasoning.

6. Considering that electricity is the flow of electrons from one place to another, propose an explanation for why metals are generally good conductors of electricity, whereas covalently bonded compounds are not.

Exercises

1. Which of the following are metallic? $F_2(g)$; $Co(s)$; $NaCl(s)$; $H_2O(s)$; $C_6H_6(l)$(benzene) ; $Pb(s)$; $Xe(g)$

2. Using grammatically correct English sentences, describe how the nature of the bonding in $MgF_2(s)$ and $Zn(s)$ differ. (Don't just name the different types of bonding—describe how they are different.)

3. J. N. Spencer, G. M. Bodner, and L. H. Rickard, *Chemistry: Structure & Dynamics*, Third Edition, John Wiley & Sons, 2006. Chapter 5: Problems: 82, 83, 85.

ChemActivity 26

The Bond-Type Triangle

(What Type of Bonding is Present?)

Model 1: The Relationships Between Electronegativities and the Physical Properties of Compounds and Metals.

Properties of Compounds with Ionic Bonding:

- high melting points (usually >500°C)
- hard and brittle as solids
- do not conduct electricity as solids; conduct electricity when molten

Properties of Metals:

- good conductors of electricity as solids
- malleable and ductile as solids
- melting points can be low (Hg, –39°C) or high (W, 3410°C)

Properties of Compounds with Covalent Bonding:

- melting points can be low (H_2, < –252°C) or high (C, diamond, > 3000°C)
- variable hardness
- do not conduct electricity as solids or when molten

Table 1. Electronegativity, EN, parameters and melting points for selected compounds.

Compound or Metal	EN		$\overline{EN}$	ΔEN	Melting Point (°C)	Type of Bonding
	first atom	second atom				
CsF(s)	0.66	4.19	2.42	3.53	682	ionic
NaCl(s)	0.87	2.87	1.87	2.00	801	ionic
NaI(s)	0.87	2.36	1.62	1.49	661	ionic
Cs(s)	0.66	0.66	0.66	0	28	metallic
Na(s)	0.87	0.87	0.87	0	98	metallic
CuZn(s;brass)	1.8	1.6	1.7	0.2	932	metallic
F_2(g)	4.19	4.19	4.19	0	–220	covalent
CH_4(g)	2.54	2.30	2.42	0.24	–182	covalent
C(s; diamond)	2.54	2.54	2.54	0	>3000	covalent
HI(g)	2.30	2.36	2.33	0.06	–51	
GaAs(s)	1.76	2.21	1.99	0.45	1238	
Si(s)	1.92	1.92	1.92	0	1410	

$\overline{EN}$ is the average electronegativity of the two elements.
ΔEN is the difference in electronegativity (absolute) between the two elements.

Critical Thinking Questions

1. Based on the data in Table 1, what combination of ΔEN values and $\overline{EN}$ values leads to metallic bonding?

2. Based on the data in Table 1, what combination of ΔEN values and $\overline{EN}$ values leads to ionic bonding?

3. Based on the data in Table 1, what combination of ΔEN values and $\overline{EN}$ values leads to covalent bonding?

4. Verify the ΔEN value and the $\overline{EN}$ value for HI, given in Table 1. Is it possible to classify HI as metallic, ionic, or covalent bonding? Explain your reasoning.

5. Verify the ΔEN value and the $\overline{EN}$ value for GaAs, given in Table 1. Is it possible to classify GaAs as metallic, ionic, or covalent bonding? Explain your reasoning.

Model 2: The Bond-type Triangle.

A **bond-type triangle** is a chart that enables us to predict the properties of a compound based on the electronegativities of the elements that comprise the compound. The data for CsF, F_2, and Cs from Table 1 have been used to generate three points at the corners of the bond-type triangle shown in Figure 1. The bond-type triangle can be divided into regions which indicate the predominant type of bonding present in compounds. The dividing lines between regions are not absolute, but they give a general idea of the boundaries between different types of bonding.

Many compounds have properties that are intermediate between the three bond types: metallic, covalent, and ionic. Si, for example, is known as a **semiconductor**; this compound has properties which are intermediate between metallic and covalent.

Figure 1. A bond-type triangle.

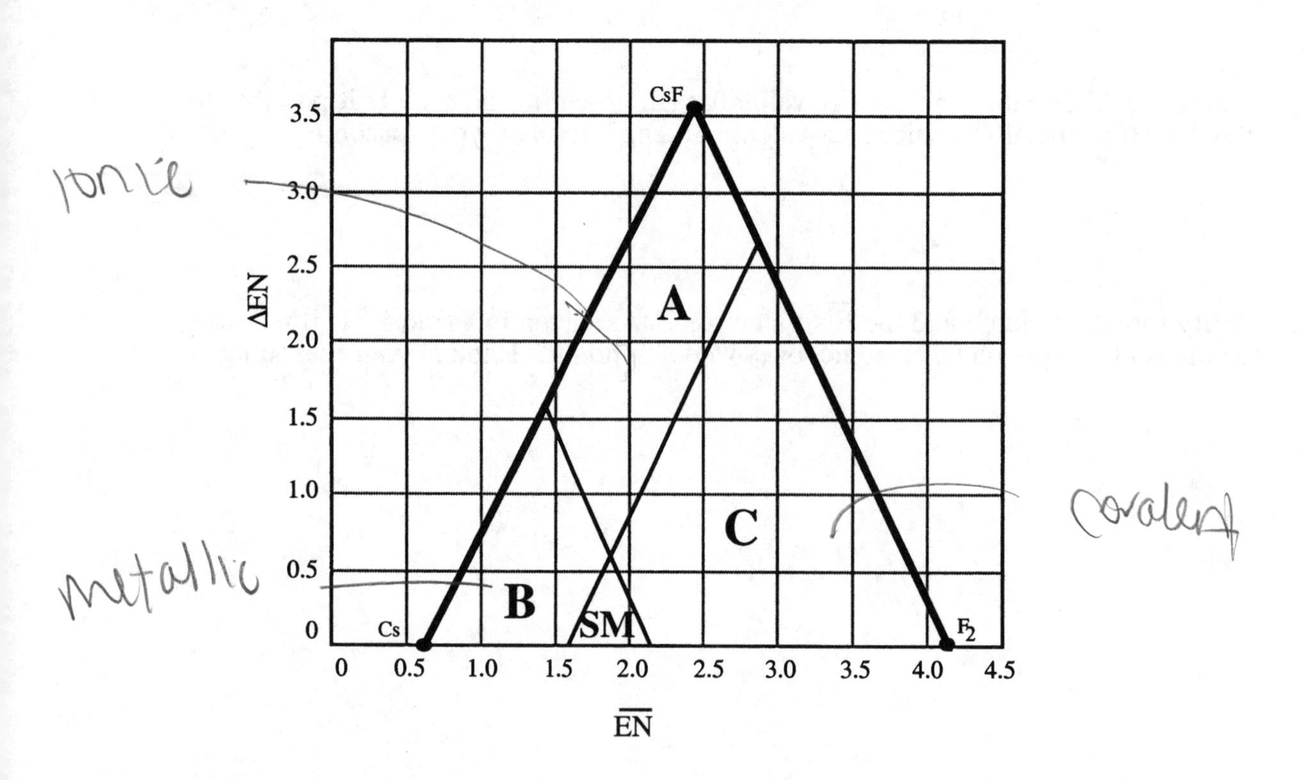

Critical Thinking Questions

6. Verify that the points for CsF, F_2, and Cs are at the appropriate positions on the bond-type triangle.

7. Place a point on the chart for sodium chloride, a compound with ionic bonding; for sodium, a metal; for methane, CH_4, a compound with covalent bonding; for Si, a semiconductor.

8. Associate the regions (A, B, C) with bond types (metallic, covalent, ionic). The "SM" region is sometimes called semimetals.

9. Quartz, SiO_2, is a very high melting, hard solid. Place a point for SiO_2 on the bond-type triangle. What type of bonding would you predict to be predominant in quartz?

Exercises

1. For each of the following compounds, place a point on the bond-type triangle. Classify each compound as metallic, covalent, ionic, semimetal.

   a) CO_2 b) NH_3 c) BaO d) SO_2 e) AlSb
   f) GaAs g) CdLi h) $BaBr_2$ i) ZnO j) NaH

2. Suggest a binary compound which would have the following characteristic [one compound for a), one compound for b), etc.]:

   a) conducts electricity in the solid state
   b) has a high melting point and is an insulator
   c) has a low melting point and is an insulator
   d) has a high melting point and conducts electricity in the solid state
   e) has a low melting point and conducts electricity in the solid state
   f) is a semiconductor

3. What type of bonding will each of the following compounds exhibit? a) A binary compound has a low $\overline{EN}$ and a low ΔEN. b) A binary compound has a high $\overline{EN}$ and a low ΔEN. a) metallic b) covalent

4. J. N. Spencer, G. M. Bodner, and L. H. Rickard, *Chemistry: Structure & Dynamics*, Third Edition, John Wiley & Sons, 2006. Chapter 5: Problems: 87, 88, 90-93, 97, 100, 101, 144, 152.

Problems

1. Answer the following question without referring to a table of electronegativities.

   In the bond-type triangle below, the position of the compound SnI_4 is indicated with an arrow. Which point (A, B, C, or D) is most likely to correspond to AgCl? Explain your reasoning clearly.

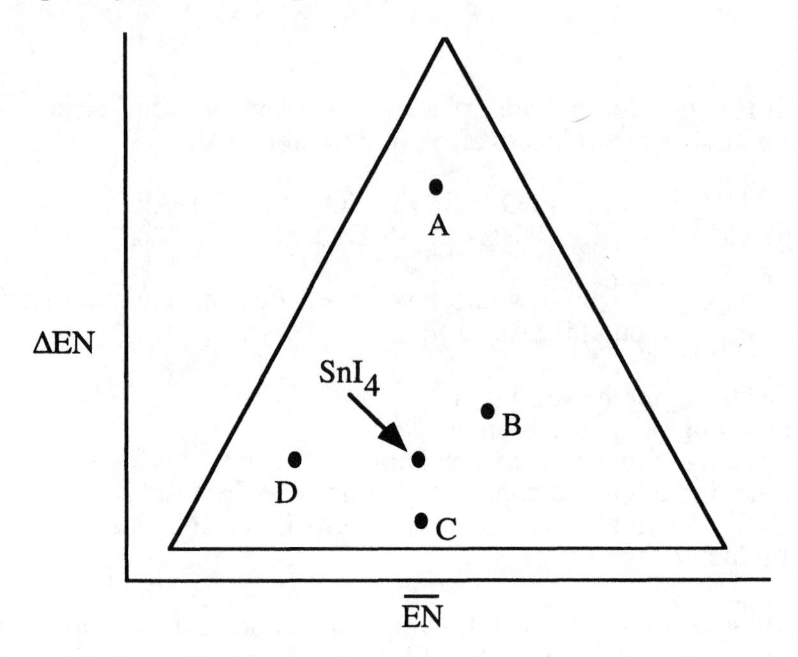

2. Give the type of bonding in each of the following: CO_3^{2-} ; $BaCO_3(s)$; $CaSO_4(s)$; $NaClO_4(s)$.

ChemActivity 27

Intermolecular Forces

(What Determines the Boiling Point?)

Model 1: Intermolecular Forces in Liquids and Gases.

Molecules attract each other, and the force of attraction increases rapidly as the intermolecular distance decreases. In a liquid, the molecules are very close to one another and are constantly moving and colliding. When a liquid evaporates, molecules in the liquid must overcome these intermolecular attractive forces and break free into the gas phase, where on average molecules are very far apart. For example, when water evaporates, rapidly moving H_2O molecules at the surface of the liquid pull away from neighboring H_2O molecules and enter the gas phase, as shown in Figure 1.

Figure 1. H_2O molecules in the liquid and gas phases.

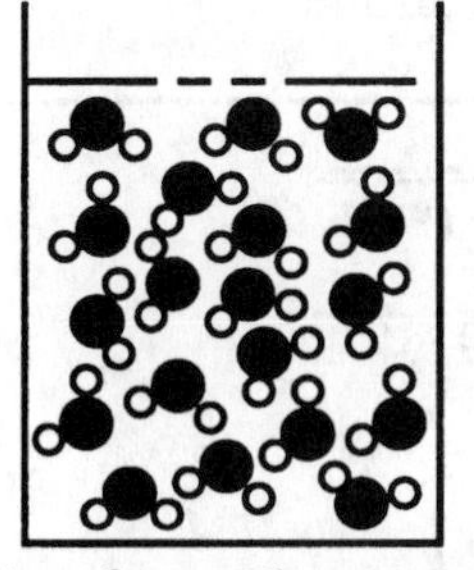

Attractive forces between water molecules are stronger in the liquid than in the gas because the molecules are very close to one another.

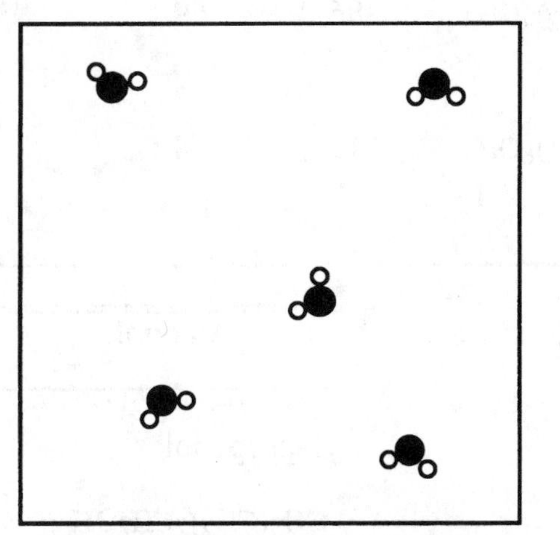

Gaseous H_2O molecules are moving fast enough to overcome the attractive forces that exist in the liquid.

Critical Thinking Questions

1. When water evaporates, are any bonds between H atoms and O atoms within a molecule broken?

2. On average, are the intermolecular forces stronger in $H_2O(l)$ or in $H_2O(g)$? Explain.

Model 2: Intermolecular Forces and Boiling Points.

To a large extent, the boiling point of a liquid is determined by the strength of the intermolecular interactions in the liquid. These interactions are largely determined by the structure of the individual molecules.

Table 1. Boiling points of selected compounds.

Alkane	MW (g/mole)	bp (°C)	Ketone	MW (g/mole)	bp (°C)
$CH_3CH_2CH_3$ propane	44.1	–42.1	$CH_3C(=O)CH_3$ acetone	58.1	56.2
$CH_3(CH_2)_2CH_3$ butane	58.1	–0.5	$CH_3C(=O)CH_2CH_3$ 2-butanone	72.1	79.6
$CH_3(CH_2)_3CH_3$ pentane	72.2	36.1	$CH_3C(=O)(CH_2)_2CH_3$ 2-pentanone	86.1	102
$CH_3(CH_2)_4CH_3$ hexane	86.2	69	$CH_3C(=O)(CH_2)_3CH_3$ 2-hexanone	100	128
$CH_3(CH_2)_8CH_3$ decane	142	174	$CH_3C(=O)(CH_2)_7CH_3$ 2-decanone	156	210

Alcohol	MW (g/mole)	bp (°C)
$CH_3CH_2CH_2OH$ 1-propanol	60.1	97.4
$CH_3(CH_2)_2CH_2OH$ 1-butanol	74.1	117
$CH_3(CH_2)_3CH_2OH$ 1-pentanol	88.2	137
$CH_3(CH_2)_4CH_2OH$ 1-hexanol	102	158
$CH_3(CH_2)_8CH_2OH$ 1-decanol	158	229

MW = Molecular Weight
Alkanes are hydrocarbons containing only C and H and have all single bonds.
Ketones contain a C=O group.
Alcohols contain an O–H group.

Critical Thinking Questions

3. Recall that the electronegativity of C and H are roughly the same, but that O has a significantly higher electronegativity. For each type of compound (alkane, ketone, alcohol) predict whether or not the compound is expected to be polar or nonpolar.

4. For each type of compound below, indicate how the boiling point changes as the molecular weight of the compound increases:

   a) alkane

   b) ketone

   c) alcohol

5. Based on your answers to CTQ 4, how do the intermolecular forces between molecules change as the molecular weight increases?

6. Find an alkane, a ketone, and an alcohol with roughly the same MW (within 5 g/mole). Rank these compounds in terms of relative boiling points.

7. a) Repeat CTQ 6 with at least two more sets of compounds.

   b) Using grammatically correct English sentences, describe any general pattern that you can identify about the relative boiling points of alkanes, ketones, and alcohols of roughly equal MW.

8. Rank the three types of compounds in terms of their relative strength of intermolecular interaction, for molecules of roughly equal MW.

9. Based on the data in Table 1, does the presence of a dipole moment in a molecule tend to increase or decrease the strength of intermolecular interactions? Explain your reasoning.

10. Is the strength of intermolecular forces determined by the bond strengths within the individual molecules? Explain your reasoning.

Model 3: Intermolecular Forces are Weaker than Covalent Bonds.

The intermolecular forces that attract molecules to each other are much weaker than the bonds that hold molecules together. For example, 463 kJ/mole are required to break one mole of O–H bonds in H_2O molecules, but only 44 kJ/mole are needed to separate one mole of water molecules in liquid water.

Read about the various types of intermolecular forces present in liquids and solids in your text or as provided by your instructor. An excellent discussion can be found in J. N. Spencer, G. M. Bodner, and L. H. Rickard, *Chemistry: Structure & Dynamics*, Third Edition, John Wiley & Sons, 2006, Sections 8.2, 8.3, 8.9, and 8.10.

Critical Thinking Questions

11. What is the difference between intramolecular bonds and intermolecular forces?

12. Rank these intermolecular forces in terms of their typical relative strengths: hydrogen bonding; dipole-dipole; induced dipole-induced dipole.

13. In the alkanes:

    a) what type(s) of intermolecular force is (are) present?

    b) what is the strongest intermolecular force present?

14. In the ketones:

    a) what type(s) of intermolecular force is (are) present?

    b) what is the strongest intermolecular force present?

15. In the alcohols:

    a) what type(s) of intermolecular force is (are) present?

    b) what is the strongest intermolecular force present?

16. In terms of intermolecular forces, why does the boiling point of a particular type of compound increase as the molecular weight increases?

17. In terms of intermolecular forces, explain the general trend that you described in CTQ 8.

Exercises

1. Based on the data in Table 1, predict the boiling points of

   a) heptane, $CH_3(CH_2)_5CH_3$
   b) ethanol, CH_3CH_2OH
   c) 2-octanone, $CH_3C(=O)(CH_2)_5CH_3$

2. Both *cis*-1,2,-dichloroethylene and *trans*-1,2,-dichloroethylene have the same molecular formula: $C_2H_2Cl_2$. However, the *cis* compound has a dipole moment, while the *trans* compound does not. One of these species has a boiling point of 60.3 °C and the other has a boiling point of 47.5 °C. Which compound has which boiling point? Explain your reasoning.

3. Rank each of the following groups of substances in order of increasing boiling point, and explain your reasoning:

   a) NH_3, He, CH_3F, CH_4
   b) CH_3Br, Ne, CH_3OH, CH_3CN
   c) CH_4, SiH_4, GeH_4, SnH_4

4. Using grammatically correct English sentences, describe the difference between the hydrogen bond between two water molecules and the O–H bond in a particular water molecule.

5. Fluoromethane, CH_3F, and methanol, CH_3OH, have approximately the same molecular weight. However, the boiling point of CH_3OH is 65.15 °C, whereas the boiling point of CH_3F is almost 100 degrees lower, –78.4 °C. Explain.

6. J. N. Spencer, G. M. Bodner, and L. H. Rickard, *Chemistry: Structure & Dynamics*, Third Edition, John Wiley & Sons, 2006. Chapter 8: Problems: 14, 55, 58, 84, 85.

Problems

1. You are given an unknown liquid to identify. You are told that the molecular formula of the compound is $C_2H_6O_2$. You measure the boiling point of the compound and find it to be 198°C. Identify this unknown liquid and explain your reasoning. You may wish to consider the following boiling points of various molecules in your analysis:

Molecule	bp (°C)
CH_4 methane	–182
CH_3CH_3 ethane	–89
$CH_3CH_2CH_3$ propane	–42
CH_3OH methanol	65
CH_3CH_2OH ethanol	78.5
$CH_3CH_2CH_2OH$ 1-propanol	97
$CH_3CH_2OCH_2CH_3$ diethyl ether	34.5

2. In each of the following groups of substances, indicate which has the highest boiling point and explain your answer.

   a) $CH_3CH_2CH_2CH_3$; $CH_3OCH_2CH_3$; $CH_3CH_2CH_2F$; $N(CH_3)_3$; $CH_3CH_2CH_2NH_2$
   b) HCl ; H_3CCF_3 ; H_2O ; CCl_4 ; NaCl
   c) LiF, O_2 ; H_2O ; CaO; I_2

3. Which of the following liquids does not exhibit hydrogen bonding?

   H_2O ; $CH_3C(=O)CH_3$; $CH_3C(=O)OH$; CH_3OH ; all of these compounds have hydrogen bonding

4. Which is the hardest to break?

   a) the H–O bond in water
   b) the hydrogen bond represented by "····" in water:

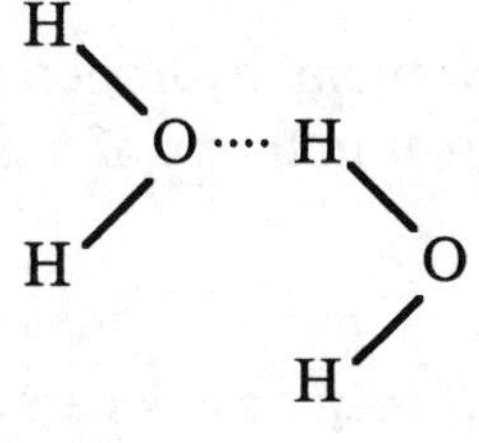

   c) the H–F bond in hydrogen fluoride
   d) the H–C bond in methane
   e) the H–N bond in ammonia

ChemActivity 28

The Mole Concept

Model: The Elephant and the Methane Molecule

One elephant has one trunk and four legs.

One methane molecule, CH_4, contains one carbon atom and four hydrogen atoms.

$$1 \text{ amu} = 1.6606 \times 10^{-24} \text{ g}$$

$$1 \text{ dozen} = 12 \text{ items}$$

$$1 \text{ mole} = 6.022 \times 10^{23} \text{ items} = \text{Avogadro's Number}$$

Critical Thinking Questions

1. How many trunks are found in one dozen elephants? Give your answer in terms of a number (such as 17 or 3.25×10^{15} trunks).

2. How many legs are found in one dozen elephants? Give your answer in terms of a number (such as 17 or 3.25×10^{15} legs).

3. How many carbon atoms are found in one dozen methane, CH_4, molecules? Give your answer in terms of a number (such as 17 or 3.25×10^{15} C atoms).

4. How many hydrogen atoms are found in one dozen methane molecules? Give your answer in terms of a number (such as 17 or 3.25×10^{15} H atoms).

5. How many trunks are found in one mole of elephants?

6. How many legs are found in one mole of elephants?

7. How many carbon atoms are found in one mole of methane molecules?

8. How many hydrogen atoms are found in one mole of methane molecules?

9. Calculate the average mass (in amu) of one methane molecule (to 0.01 amu).

10. Based on your answer to CTQ 9, calculate the mass (in grams) of one mole of methane molecules (to 0.01 g).

11. Use a grammatically correct English sentence to describe how the mass in amu of one molecule of a compound is related to the mass in grams of one mole of that compound.

Exercises

Unless otherwise stated, calculate all mass values in grams.

1. What is the mass of 1.00 mole of Cu?

2. What is the mass of 1.00 mole of sodium(I) fluoride, NaF?

3. Consider a 1.00 carat diamond (pure C) that has a mass of 0.200 grams. How many carbon atoms are present in this diamond? Give your answer in terms of a number (such as 17 or 3.25×10^{15} C atoms) and as a number of moles of C atoms.

4. Consider 1.00 mole of dihydrogen gas, H_2. How many dihydrogen molecules are present? How many hydrogen atoms are present? What is the mass of this sample?

5. Ethanol has a molecular formula of CH_3CH_2OH. What is the mass of 1.000 moles of ethanol? What is the average mass of one molecule of ethanol?

6. What is the mass of 0.5623 moles of ethanol, CH_3CH_2OH?

7. a) How many moles of ethanol are present in a 100.0 g sample of ethanol?

   b) How many moles of each element (C, H, O) are present in a 100.0 g sample of ethanol?

c) How many grams of each element (C, H, O) are present in a 100.0 g sample of ethanol?

8. How many moles of carbon dioxide, CO_2, are present in a sample of carbon dioxide with a mass of 254 grams?

9. How many moles of O atoms are present in a 254 g sample of carbon dioxide?

10. How many carbon atoms are found in 0.500 g of glycine, H_2NCH_2COOH?

11. Indicate whether each of the following statements is true or false, and explain your reasoning.

    a) One mole of NH_3 weighs more than one mole of H_2O.
    b) There are more carbon atoms in 48 grams of CO_2 than in 12 grams of diamond (a form of pure carbon).
    c) There are equal numbers of nitrogen atoms in one mole of NH_3 and one mole of N_2.
    d) The number of Cu atoms in 100 grams of Cu(s) is the same as the number of Cu atoms in 100 grams of copper(II) oxide, CuO.
    e) The number of Ni atoms in 100 moles of Ni(s) is the same as the number of Ni atoms in 100 moles of nickel(II) chloride, $NiCl_2$.
    f) There are more hydrogen atoms in 2 moles of NH_3 than in 2 moles of CH_4.

12. Use grammatically correct English sentences to describe how to calculate the number of H atoms in "z" moles of NH_3.

13. J. N. Spencer, G. M. Bodner, and L. H. Rickard, *Chemistry: Structure & Dynamics*, Third Edition, John Wiley & Sons, 2006. Chapter 1: Problems: 122-125, 127, 129, 131, 132, 140, 141.

ChemActivity 29

Balanced Chemical Equations

(What Happens When a Chemical Reaction Occurs?)

Model 1: Two Balanced Chemical Reactions.

Two balanced chemical reactions (or equations) are given below:

$$CuO(s) + H_2(g) \rightarrow Cu(s) + H_2O(g) \quad (1)$$

$$2CO(g) + O_2(g) \rightarrow 2CO_2(g) \quad (2)$$

Critical Thinking Questions

1. Indicate the reactants and products for each reaction in the table below:

Reaction	Reactant(s)	Product(s)
(1)		
(2)		

2. What does the arrow represent in a chemical reaction?

3. For reaction (1), how many H atoms, Cu atoms, and O atoms are represented on:

   a) the reactant side?

   b) the product side?

4. For reaction (2), how many C atoms and O atoms are represented on:

   a) the reactant side?

   b) the product side?

5. Based on your answers to CTQs 3 and 4, what general statement can be made about the number of atoms of each type on the two sides of a balanced chemical equation?

Information

Atoms are neither created nor destroyed when chemical reactions take place. Therefore, the number of atoms of each element must be identical on the reactant (left) and product (right) sides of a balanced chemical reaction. Such a chemical equation is said to be **atom balanced.**

Model 2: Four Balanced Chemical Reactions.

In each of the balanced chemical reactions given below, the symbol "(aq)" indicates that the molecule or ion is surrounded by water molecules.

$$Ag^{+}(aq) + Cl^{-}(aq) \rightarrow AgCl(s) \qquad (3)$$

$$Zn(s) + Cu^{2+}(aq) \rightarrow Zn^{2+}(aq) + Cu(s) \qquad (4)$$

$$3ClO^{-}(aq) \rightarrow 2Cl^{-}(aq) + ClO_3^{-}(aq) \qquad (5)$$

$$2Cr^{2+}(aq) + Mg^{2+}(aq) \rightarrow 2Cr^{3+}(aq) + Mg(s) \qquad (6)$$

Critical Thinking Questions

6. Confirm that each of the chemical equations in Model 2 are *atom balanced.*

7. a) For each of the chemical equations in Model 2, determine the sum of the charges on the left-hand side and the sum of the charges on the right-hand side.

   b) Based on the reactions in Model 2, which, if any, of the following statements are correct?

      i) The sum of the charges on both sides of a balanced chemical equation must equal zero.

      ii) The sum of the charges on both sides of a balanced chemical equation must be a positive number.

      iii) The sum of the charges on both sides of a balanced chemical equation must be a negative number.

8. What general statement can be made about the sum of the charges on both sides of a a balanced chemical equation?

Information

Protons and electrons are neither created nor destroyed when chemical reactions take place. Therefore, the total charge must be identical on the reactant and product sides of a balanced chemical equation. Such a chemical equation is said to be **charge balanced**.

Exercises

1. Balance these chemical reactions:

   a) $Cr(s) + S_8(s) \rightarrow Cr_2S_3(s)$
   b) $NaHCO_3(s) \rightarrow Na_2CO_3(s) + CO_2(g) + H_2O(g)$
   c) $Fe_2S_3(s) + HCl(g) \rightarrow FeCl_3(s) + H_2S(g)$
   d) $CS_2(l) + NH_3(g) \rightarrow H_2S(g) + NH_4SCN(s)$

2. Write a balanced chemical equation for the gaseous reaction of methane (CH_4) with oxygen (O_2) to form carbon dioxide (CO_2) and water (H_2O).

3. Write a chemical equation that forms one mole of glycine, $H_2NCH_2COOH(s)$, from solid carbon, gaseous oxygen, gaseous nitrogen, and gaseous hydrogen.

4. Write a chemical equation that has only ozone, O_3, on the left-hand side and only molecular oxygen on the right-hand side.

5. Which of the following chemical equations are not balanced?

   a) $NO_2^-(aq) + ClO_2^-(aq) \rightarrow NO_3^-(aq) + Cl^-(aq)$
   b) $NO_2^-(aq) + ClO^-(aq) \rightarrow NO_3^-(aq) + Cl^-(aq)$
   c) $Cr(s) + Pb^{2+}(aq) \rightarrow Pb(s) + Cr^{3+}(aq)$
   d) $H^+(aq) + SO_3^{2-} \rightarrow HSO_3^-(aq)$
   e) $4AgBr(s) + 4OH^-(aq) \rightarrow O_2(g) + 2H_2O + 4Ag(s) + 4Br^-(aq)$

6. J. N. Spencer, G. M. Bodner, and L. H. Rickard, *Chemistry: Structure & Dynamics*, Third Edition, John Wiley & Sons, 2006. Chapter 2: Problems: 13-17, 88.

Model 3: The Balanced Chemical Equation.

A balanced chemical reaction can be interpreted in two ways. First, it can be thought of as describing how many molecules of reactants are consumed in order to produce a certain number of molecules of products. Analogously, it can be thought of as describing how many *moles* of reactants are consumed in order to produce the indicated number of *moles* of products.

$$CuO(s) + H_2(g) \rightarrow Cu(s) + H_2O(g) \quad (1)$$

$$2CO(g) + O_2(g) \rightarrow 2CO_2(g) \quad (2)$$

Critical Thinking Questions

9. How many H_2O molecules are produced for every H_2 molecule that is consumed in reaction (1)?

10. For reaction (2):

    a) How many CO_2 molecules are produced for every O_2 molecule consumed?

    b) How many CO_2 molecules are produced for every CO molecule consumed?

    c) How many molecules of CO_2 are produced when 2 molecules of O_2 are consumed?

    d) How many moles of CO_2 are produced when 5 moles of O_2 are consumed?

11. How many moles of CuO react in order to produce 12 moles of Cu in reaction (1)?

12. Determine the number of reactant molecules and the number of product molecules for reaction (1) and reaction (2).

13. a) Is the number of molecules identical on the reactant and product sides of these balanced equations?

    b) Does the total number of moles of gas increase, decrease, or remain constant when reaction (2) occurs?

14. Explain how your answers to CTQ 13 can be consistent with the idea that atoms are neither created nor destroyed when chemical reactions take place.

15. Is it correct to state that if 100 grams of CuO are consumed when reaction (1) occurs, then 100 g of Cu are formed in the process? Why or why not?

16. Describe, using grammatically correct English sentences, the steps taken to calculate the number of grams of CO_2 produced in reaction (2) given that X grams of O_2 are consumed.

Exercises

7. How many grams of Cr_2S_3 are produced when the reaction in Ex. 1a (above) occurs with 10.0 grams of chromium being consumed?

8. How many grams of hydrogen sulfide are produced when 0.0365 grams of carbon disulfide are consumed in the reaction in Ex. 1d?

9. How many grams of iron(III) chloride are produced when 26 grams of hydrogen sulfide gas are produced in the reaction in Ex. 1c?

10. The thermite reaction has been used for welding railroad rails, in incendiary bombs, and to ignite solid-fuel rockets. The reaction is

$$Fe_2O_3(s) + 2Al(s) \rightarrow 2Fe(l) + Al_2O_3(s)$$

What masses of iron(III) oxide and aluminum must be used to produce 15.0 g of iron? What is the mass of aluminum oxide that would be produced?

11. Nitrogen (N_2) combines with hydrogen (H_2) to form ammonia (NH_3). How many grams of ammonia are formed when 145 grams of nitrogen are consumed by hydrogen?

12. Indicate whether the following statement is true or false and explain your reasoning. When carbon monoxide gas reacts with oxygen gas to form carbon dioxide gas, the number of gas molecules present decreases.

13. J. N. Spencer, G. M. Bodner, and L. H. Rickard, *Chemistry: Structure & Dynamics*, Third Edition, John Wiley & Sons, 2006. Chapter 2: Problems: 21-23, 25, 27, 30, 32, 33.

Problems

1. Nickel can react with gaseous carbon monoxide to form $Ni(CO)_4$. Other metals present do not react. If 94.2 grams of a mixture of metals reacts with carbon monoxide to produce 98.4 grams of $Ni(CO)_4$, what is the mass percent of nickel in the original sample?

2. A 1.000 g sample of iron reacts with element "Q" to form 1.430 g of Fe_2Q_3. a) Determine the identity of element "Q". b) Write a balanced chemical equation for this reaction.

ChemActivity 30

Limiting Reagent

(How Much Can You Make?)

Model 1: The S'more.

A delicious treat known as a S'more is constructed with the following ingredients and amounts:

1 graham cracker
1 chocolate bar
2 marshmallows

At a particular store, these items can be obtained only in full boxes, each of which contains one gross of items. A gross is a specific number of items, analogous (but not equal) to one dozen. The boxes of items have the following net weights (the weight of the material inside the box):

box of graham crackers	9.0 pounds
box of chocolate bars	36.0 pounds
box of marshmallows	3.0 pounds

Critical Thinking Questions

1. If you have a collection of 100 graham crackers, how many chocolate bars and how many marshmallows do you need to make S'mores with all of the graham crackers?

2. If you have a collection of 1000 graham crackers, 800 chocolate bars, and 1000 marshmallows:

   a) How many S'mores can you make?

   b) What (if anything) will be left over, and how many of that item will there be?

Information

Chemists refer to the reactant which limits the amount of product that can be made from a given collection of original reagents as the **limiting reagent** or **limiting reactant**.

Critical Thinking Questions

3. Identify the limiting reagent for CTQ 2.

4. Based on the information given, which of the three ingredients (a graham cracker, a chocolate bar, or a marshmallow):

   a) weighs the most?

   b) weighs the least?

   Explain your reasoning.

5. If you have 36.0 pounds of graham crackers, 36.0 pounds of chocolate bars, and 36.0 pounds of marshmallows:

   a) which item do you have the most of?

   b) which item do you have the least of?

   Explain your reasoning.

6. a) If you attempt to make S'mores from the material described in CTQ 5, what will be the limiting reagent?

   b) How many gross of S'mores can you make?

   c) How many gross of each of the two leftover items will you have?

   d) How many pounds of each of the leftover items will you have?

   e) How many pounds of S'mores will you have?

7. Using G as the symbol for graham cracker, Ch for chocolate bar, and M for marshmallow, write a "balanced chemical equation" for the production of S'mores.

8. Using grammatically correct English sentences, explain why is it not correct to state that if we start with 36 pounds each of G, Ch, and M, then we should end up with $3 \times 36 = 108$ pounds of S'mores.

9. Given the "balanced chemical equation" for the production of S'mores from CTQ 7, calculate the mass of S'mores that can be made from 416 pounds of chocolate bars, 142 pounds of graham crackers, and 58.2 pounds of marshmallows.

Exercises

1. Given the balanced chemical equation:

   $$2\,NO(g) + O_2(g) \rightarrow 2\,NO_2(g)$$

   calculate the mass of nitrogen dioxide that can be made from 30.0 grams of NO and 30.0 grams of O_2.

2. Zinc, Zn, and iodine, I_2, react to form zinc(II) iodide, ZnI_2 (the reactants and the product are all solids at room temperature).

    a) Write a balanced chemical equation for this reaction.
    b) Suppose that 50.0 g of zinc and 50.0 g of iodine are used to form zinc(II) iodide.
        1) Assuming that the reaction goes to completion, which element will be totally consumed in the formation of the zinc(II) iodide?
        2) What is the limiting reagent?
        3) How many grams of zinc(II) iodide can be produced?
        4) How many grams of the excess element remain unreacted?

3. Acetylene gas, HCCH, is commonly used in high temperature torches.

    a) Write a balanced chemical equation for the reaction of acetylene with hydrogen gas (H_2) to form ethane (C_2H_6).
    b) How many grams of ethane can be produced from a mixture of 30.3 grams of HCCH and 4.14 grams of H_2?

4. Titanium (Ti) is a strong, lightweight metal that is used in the construction of rockets, jet engines, and bicycles. It can be prepared by reacting $TiCl_4$ with Mg metal at very high temperatures. The products are Ti(s) and $MgCl_2$.

    a) Provide a balanced chemical reaction for the reaction described above.
    b) How many grams of Ti metal can be produced from a reaction involving 3.54×10^4 g of $TiCl_4$ and 6.53×10^3 g of Mg?

5. The first step in the manufacturing process of phosphorous is the reaction below:

    $$2\ Ca_3(PO_4)_2(s) + 6\ SiO_2(s) \rightarrow 6\ CaSiO_3(s) + P_4O_{10}(g)$$

    The MW of $Ca_3(PO_4)_2(s)$ is 310.2 g/mole and the MW of $SiO_2(s)$ is 60.1 g/mole. If 20.0 g of $Ca_3(PO_4)_2(s)$ and 20.0 g of $SiO_2(s)$ are reacted, how many grams of $P_4O_{10}(g)$ can be produced?

6. How many grams of N_2 (28.01 g/mole) can be obtained by reacting 24.5 g of NH_3 (17.03 g/mole) with 30.8 g of O_2 (MW = 32.00 g/mole)?

    $$4\ NH_3(g) + 3\ O_2(g) \rightarrow 2\ N_2(g) + 6\ H_2O(l)$$

7. J. N. Spencer, G. M. Bodner, and L. H. Rickard, *Chemistry: Structure & Dynamics*, Third Edition, John Wiley & Sons, 2006. Chapter 2: Problems: 35-40, 42, 87.

ChemActivity 31

Empirical Formula

(Can a Molecule Be Identified by Its Percent Composition?)

Model: Percent Composition.

The **percent composition** (by mass) of an element in a molecule is the mass of the element in the molecule divided by the mass of the entire molecule times 100. Or, because the number of atoms (molecules) is proportional to the number of moles of atoms (molecules),

$$\text{percent composition of element } i = \frac{\text{mass of } i \text{ in one mole of the compound}}{\text{mass of one mole of the compound}} \times 100\%$$

Table 1. Percent composition (by mass) of some common organic molecules.

Name	Structural Formula	Molecular Formula	% Composition (by mass) C	H
ethyne	HC≡CH	C_2H_2	92.26	7.74
benzene	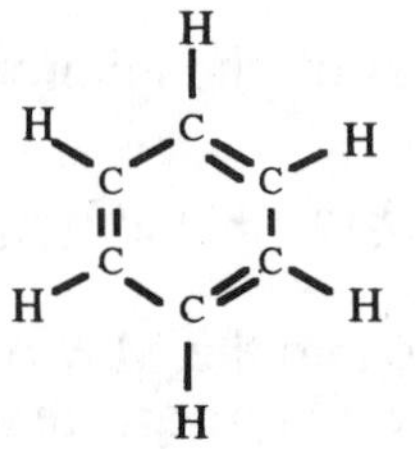			
cyclobutane	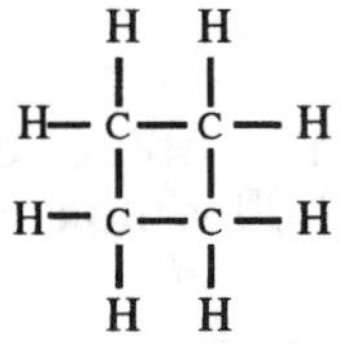	C_4H_8		
2-butene	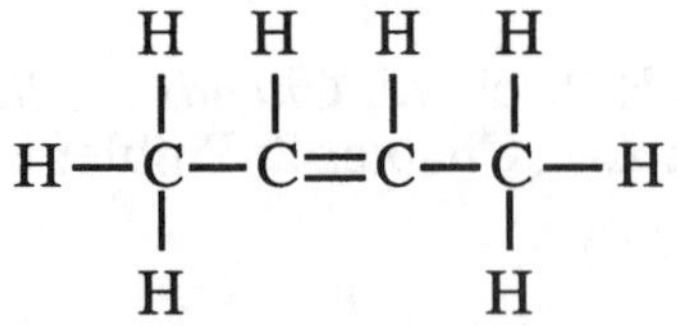		85.63	
1-octene	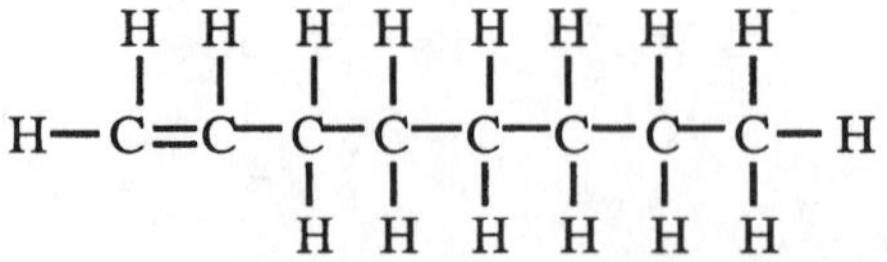			

Critical Thinking Questions

1. Verify that the % composition given for ethyne in Table 1 is correct.

2. Fill in the missing molecular formulas and % compositions in Table 1.

3. Is it possible, given the original data in Table 1, to determine the % composition by mass of H for 2-butene without using the equation given in the model above? If so, how?

4. Based on the data in Table 1, is it possible to determine the *molecular* formula of a compound solely from its percent composition? Why or why not?

5. What feature related to composition do all compounds with the same % composition have?

Information

The **empirical formula** of a compound describes the relative number of each type of atom in the compound. It is given in terms of the smallest-possible-whole-number ratios (as subscripts). For example, the empirical formula of ethane is CH_3. (Note that the subscript "1" is omitted.)

Critical Thinking Questions

6. What feature related to the composition of a compound can be determined solely by percent composition?

7. Determine the empirical formula of each of the molecules in Table 1.

Exercises

1. The molecule 2-hexene has the molecular formula C_6H_{12}. Refer to Table 1 and determine the percent composition of H in this molecule.

2. Determine the percent composition of each element in acetic acid, CH_3COOH.

3. A molecule containing only nitrogen and oxygen contains (by mass) 36.8% N.

   a) How many grams of N would be found in a 100 g sample of the compound? How many grams of O would be found in the same sample?
   b) How many moles of N would be found in a 100 g sample of the compound? How many moles of O would be found in the same sample?
   c) What is the ratio of the number of moles of O to the number of moles of N?
   d) What is the empirical formula of the compound?

4. A compound used as a dry-cleaning fluid was analyzed and found to contain 18.00% C, 2.27% H, and 79.73% Cl. Determine the empirical formula of the fluid.

5. J. N. Spencer, G. M. Bodner, and L. H. Rickard, *Chemistry: Structure & Dynamics*, Third Edition, John Wiley & Sons, 2006. Chapter 1: Problems: 135, 137, 139, 145, 146, 149, 150, 152, 153, 155, 157, 165, 166, 169, 171.

Problems

1. Indicate whether the following statement is true or false and explain your reasoning.

   All compounds with the same empirical formula are isomers of each other.

2. An unknown liquid contains 38.7% C and 51.6% O by mass. The remainder of the compound is H. What is the empirical formula of the compound?

3. A compound containing only P, O, and Zn is used as a dental cement. A sample of the cement is analyzed and gives 33.16% O and 16.04% P. Determine the empirical formula of the cement.

ChemActivity 32

Molarity

(How Concentrated Is It?)

Water is the most common solvent, and we will focus on aqueous solutions. However, the terms applied herein also apply when other solvents are used.

Solutes are categorized according to their ability to affect the electrical conductivity of the solution upon dissolution of the solute in water.

- An aqueous solution of a **strong electrolyte** conducts electricity well.
- An aqueous solution of a **nonelectrolyte** does not conduct electricity.
- An aqueous solution of a **weak electrolyte** conducts electricity poorly.
- Water from most common sources (tap water, rain water, water in lakes and rivers, etc.) always contains strong electrolytes and conducts electricity well.

Model 1: Strong Electrolytes.

Many ionic compounds are strong electrolytes. When an ionic compound dissolves in water the cations and anions are separated. In the solution, the cations and anions are surrounded by water molecules. Sodium sulfate and sodium chloride are both strong electrolytes.

$$Na_2SO_4(s) \xrightarrow{H_2O} 2\,Na^+(aq) + SO_4^{2-}(aq)$$

$$NaCl(s) \xrightarrow{H_2O} Na^+(aq) + Cl^-{}_{(aq)}$$

The "(aq)" after the ion indicates that each ion of that type is surrounded by several water molecules. It has been discovered that it is the mobile, charged particles in the solution that carry the electric current.

Figure 1: Cations and anions in a solid crystal and in water.

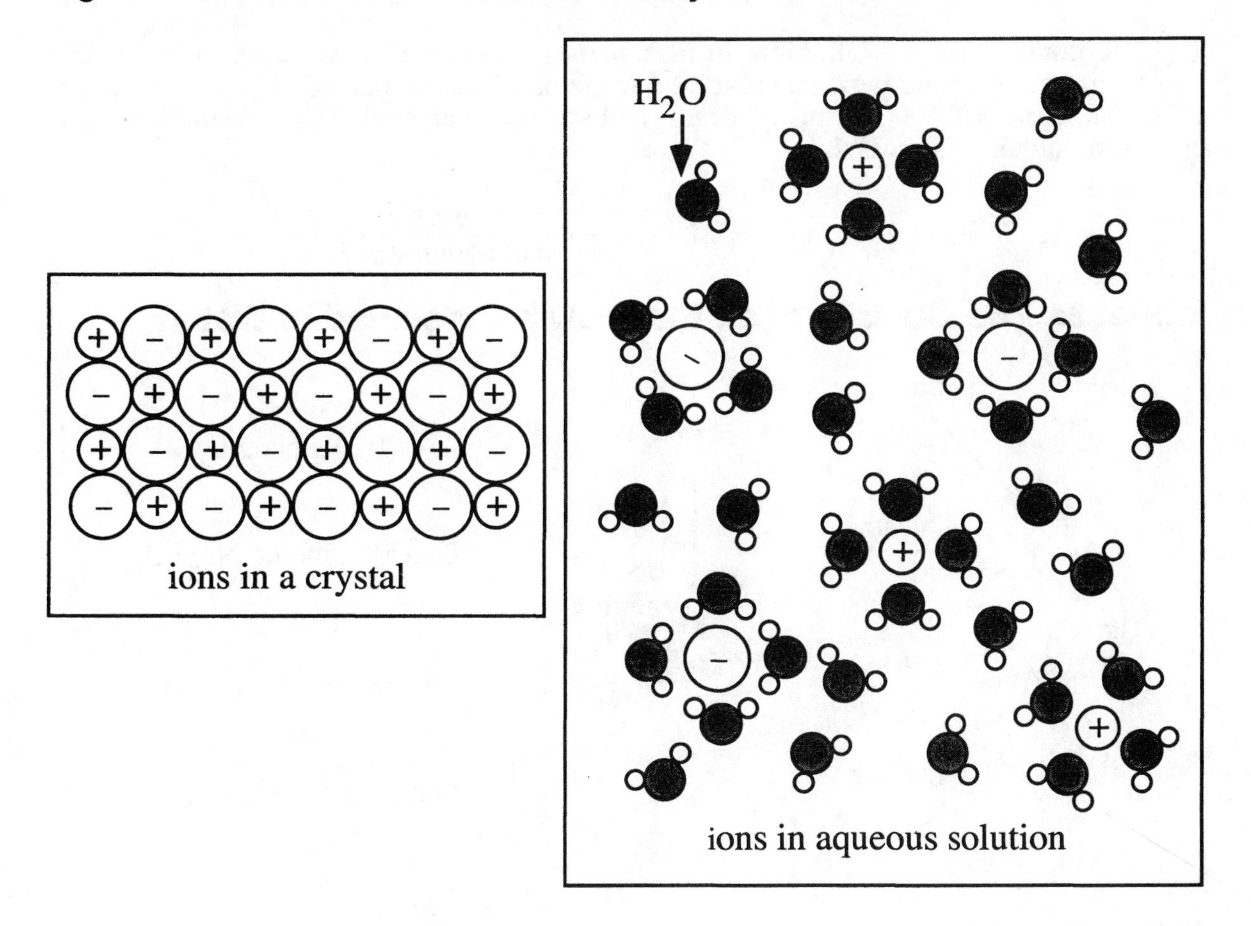

Critical Thinking Questions

1. Which ionic compound, NaCl or Na_2SO_4, does Figure 1 represent? Explain your reasoning clearly.

2. a) What is the identity of the cation in Figure 1?

   b) What is the identity of the anion in Figure 1?

3. Which atom(s) of the water molecule is closest to the cations in solution? Why?

4. Which atom(s) of the water molecule is closest to the anions in solution? Why?

Information

The concentration of a solute in an aqueous solution can be expressed in many ways—grams of solute per liter of solution; grams of solute per 1000 grams of water; moles of solute per 1000 grams of water; and so on. One of the most frequently used concentration units is **molarity**.

$$\text{Molarity of solute } i = M_i = \frac{\text{moles of } i}{\text{volume of solution in Liters}}$$

Model 2: The Molarity of an Electrolyte Dissolved in Water.

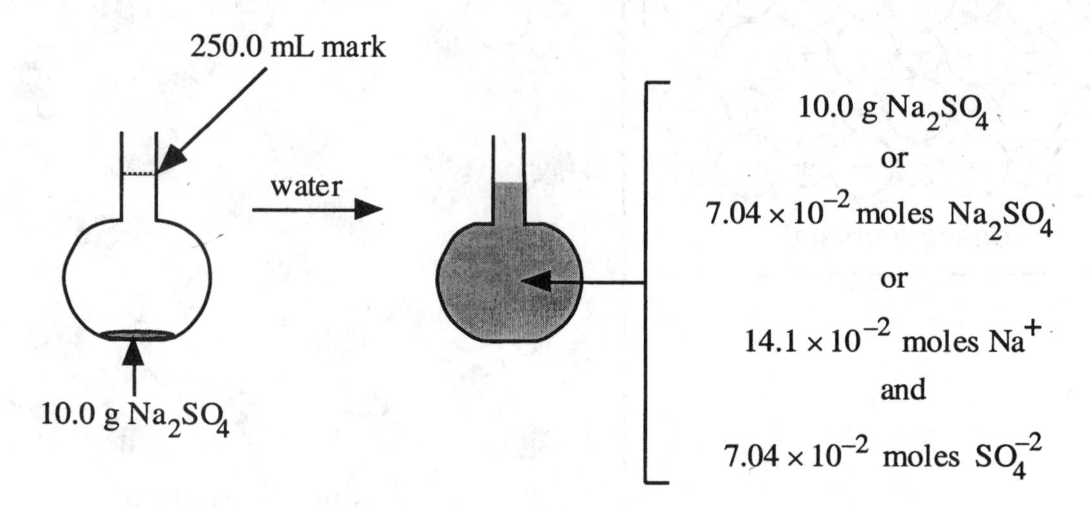

Critical Thinking Questions

5. When one mole of Na_2SO_4 is dissolved in water:

   a) how many moles of sodium ions are found in the solution?

   b) how many moles of sulfate ions are found in the solution?

6. Verify that when 10.0 g of sodium sulfate dissolves in water:

   a) there are 7.04×10^{-2} moles of sodium sulfate in the water.

b) there are 7.04×10^{-2} moles of sulfate in the water.

c) there are 14.1×10^{-2} moles of sodium in the water.

7. a) What is the molarity of the sodium sulfate solution in Model 2?

b) What is the molarity of the sulfate ions in the solution in Model 2?

c) What is the molarity of the sodium ions in the solution in Model 2?

8. Which is more concentrated with respect to sodium ions, 50.0 g of NaCl in 500.0 mL of solution or 59.0 g of Na_2SO_4 in 500.0 mL of solution?

9. Which is more concentrated with respect to sodium ions, 0.50 M NaCl or 0.30 M Na_2SO_4?

10. Which CTQ was easier to answer, 8 or 9? Why?

Information

Some molecules do not dissociate into ions when dissolved in water. Sugars (glucose, sucrose, dextrose, etc.) and alcohols are examples. These compounds do not dissociate into ions upon dissolution and they do not increase the conductivity of water. When glucose, $C_6H_{12}O_6$, dissolves in water, each glucose molecule is surrounded by water.

$$C_6H_{12}O_6(s) \xrightarrow{H_2O} C_6H_{12}O_6(aq)$$

Model 3: A Nonelectrolyte Dissolved in Water.

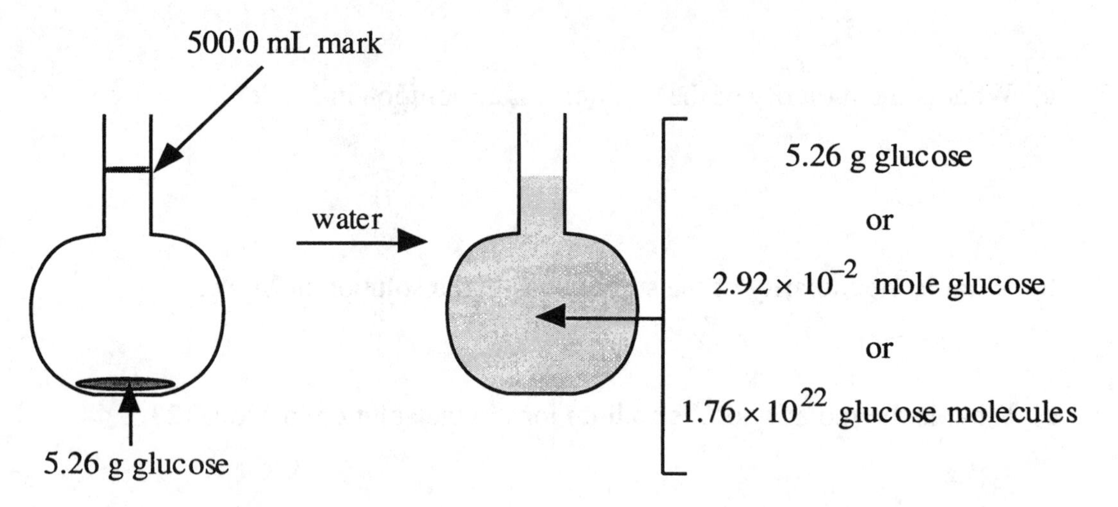

Critical Thinking Questions

11. Verify that when 5.26 g of glucose, $C_6H_{12}O_6$, dissolves in water:

    a) there are 2.92×10^{-2} moles of glucose in the water.

    b) there are 1.76×10^{22} molecules of glucose in the water.

12. What is the glucose molarity of the solution in Model 3?

Exercises

1. Determine the aluminum ion concentration and the chloride ion concentration in 0.125 M $AlCl_3$.

2. Determine the chromium ion concentration and the sulfate ion concentration in a 300 mL solution that contains 5.00 g of chromium(III) sulfate.

3. Determine the mass of metal ions in each of the following: 500 mL of 0.752 M NaCl; 750 mL of 2.54×10^{-5} M $Pb(NO_3)_2$; 10.5 L of 0.209 M Na_3PO_4 .

4. Indicate whether the following statement is true or false, and explain your reasoning:

   The number of solute particles present in 1.0 L of 0.50 M Na_2SO_4 is the same as the number of solute particles present in 1.0 L of 0.50 M sucrose.

5. A solution is prepared by dissolving 0.5482 grams of iron(III) nitrate in enough water to make 100.0 mL of solution. A 10.00-mL aliquot (portion) of this solution is then diluted to a final volume of 250.0 mL. What is the concentration of Fe^{3+} ions in the final solution?

6. AgCl is essentially insoluble in water. If a solution containing Ag^+ ions is mixed with a solution containing Cl^- ions, the following reaction occurs:

   $$Ag^+(aq) + Cl^-(aq) \rightarrow AgCl(s)$$

   producing a solid precipitate of AgCl.

   How many moles of solid AgCl can be produced when 25.0 mL of a 0.125 molar NaCl solution is mixed with 35.0 mL of a 0.100 molar $AgNO_3$ solution?

7. J. N. Spencer, G. M. Bodner, and L. H. Rickard, *Chemistry: Structure & Dynamics*, Third Edition, John Wiley & Sons, 2006. Chapter 2: Problems: 49, 50, 53, 55, 56, 77, 83.

Problems

1. A large amount of an unknown metal, M, reacts with 4.60 grams of Cl_2 to produce 6.84 grams of a pure metal chloride. When the metal chloride is dissolved in a 100.0-mL volumetric flask which is then filled up to the mark, the concentration of metal ions is found to be 0.43 moles/liter. What is the unknown metal, M? Explain your reasoning carefully.

2. Which of the following solutions has the highest concentration of chloride ions?
   a) 10.0 g of NaCl dissolved in 50.0 mL of solution.
   b) 15.0 g of $CaCl_2$ dissolved in 100.0 mL of solution.
   c) 20.0 g of $CrCl_3$ dissolved in 125.0 mL of solution.

3. Carefully describe how you would make 500 mL of 0.150 M Na_2SO_4 given 1 kg of Na_2SO_4, a distilled water supply, a balance, and a 500-mL volumetric flask.

4. Suppose that 400 mL of 0.0700 M $BaCl_2$ is added to 800 mL of 0.0300 M $BaCl_2$. Assume that the volumes are additive and calculate the chloride ion concentration in the final solution.

ChemActivity 33

The Ideal Gas Law

(How Do Gases Behave?)

Information

- T (K) = Kelvin or Absolute temperature = T(°C) + 273.15°
  T (K) is always > 0.

- Boyle's Law (1660): The volume of a sample of a gas varies inversely with pressure, if the temperature is held constant.

$$V = k_B \frac{1}{P} \qquad \text{at constant } n \text{ and } T$$

  where n is the number of moles of gas.

- Charles' Law (1887): The volume of a gas varies linearly with temperature, if the pressure is held constant.

$$V = k_C T \qquad \text{at constant } n \text{ and } P$$

- Avogadro's Hypothesis (1812): Samples of different gases which contain the same number of molecules—of any complexity, size, or shape—occupy the same volume at the same temperature and pressure.

$$V = k_A n \qquad \text{at constant } T \text{ and } P$$

Model 1: The Ideal Gas Law Equation.

$$PV = nRT$$

where R is a constant called the universal gas constant.

The numerical value of the universal gas constant is calculated from the fact that one mole of gas occupies 22.414 L at a pressure of one atmosphere and a temperature of 0°C (273.15 K).

$$R = \frac{PV}{nT} = \frac{(1 \text{ atm})(22.414 \text{ L})}{(1 \text{ mole})(273.15 \text{ K})} = 0.08206 \frac{\text{L atm}}{\text{K mol}}$$

Critical Thinking Questions

1. How does the volume of a gas (at constant n and P) change as the temperature is raised?

2. How does the volume of a gas (at constant n and T) change as the pressure is increased?

3. How does the volume of a gas (at constant T and P) change as the number of molecules is increased?

4. For each case, rearrange the Ideal Gas Law Equation to show that it is consistent with the given law or hypothesis and obtain an expression for the corresponding constant.

   a) Boyle's Law, k_B

   b) Charles' Law, k_C

   c) Avogadro's Hypothesis, k_A

Exercises

1. Calculate the volume of 20.5 g of NH_3 at 0.658 atm and 25 °C.

2. Calculate the volume of 359 g of CH_3CH_3 at 0.658 atm and 75 °C.

3. Calculate the volume of 525 g of O_2 at 25.7 torr (760 torr = 1 atm) and 25 °C.

4. A spherical space colony proposed by Gerald O'Neill (Princeton University) has a diameter of 1.00 km. How many grams of N_2 are needed to fill the interior of the colony at one atmosphere and 20 °C (room temperature)?

5. A 2.00 L container is placed in a constant temperature bath and is filled with 3.05 g of CH_3CH_3. The pressure stabilizes at 800 torr. What is the temperature of the constant temperature bath?

6. The density of a gas is typically given as: density $= d = \frac{\text{grams}}{\text{liter}}$. Use this definition of density and the ideal gas law to derive an equation that has only the density on the left-hand side and the other variables (P, T, MW) on the right-hand side.

7. Calculate the density of NH_3 at 850 torr and 100 °C.

8. J. N. Spencer, G. M. Bodner, and L. H. Rickard, *Chemistry: Structure & Dynamics*, Third Edition, John Wiley & Sons, 2006. Chapter 6: Problems: 29, 31, 33, 37, 43-45, 48, 50, 53, 56, 57, 65, 69.

Information

In a mixture of gases the total pressure, P_T, is the sum of the pressures of the individual gases, P_i.

$$P_T = \sum_i P_i \tag{1}$$

The partial pressure of each gas in the mixture is given by

$$P_i = n_i \frac{RT}{V} \tag{2}$$

Exercises

9. A 2.00 L container holds 4.00 moles of O_2 and 2.70 moles of He at 293 K. What is the partial pressure of O_2? Of He? What is the total pressure?

10. The density of air at 1.000 atm and 25 °C is 1.186 g/L.

    a) Calculate the average molecular mass of air. $\cdot\ 1.186 = \frac{mw(1)}{.08206\,(298.15)} = 29.02$ mol

    b) From this value, and assuming that air contains only molecular nitrogen and molecular oxygen gases, calculate the mass % of N_2 and O_2 in air.

11. J. N. Spencer, G. M. Bodner, and L. H. Rickard, *Chemistry: Structure & Dynamics*, Third Edition, John Wiley & Sons, 2006. Chapter 6: Problems: 72, 75, 93, 94, 105, 106, 108.

Problems

1. Consider the three flasks in the figure below. Assume that the connecting tubes have no volume and the temperature is held constant.

    a) Calculate the partial pressure of each gas when all stopcocks are open.

    b) Calculate the total pressure when all stopcocks are open.

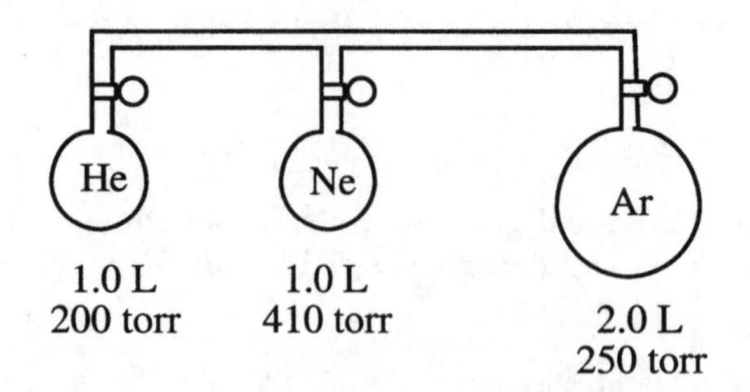

2. Some commercial drain cleaners contain two components: sodium(I) hydroxide and aluminum powder. When the mixture is poured down a clogged drain, the following reaction occurs:

$$2\ NaOH(aq) + 2\ Al(s) + 6\ H_2O(l) \rightarrow 2\ NaAl(OH)_4(aq) + 3\ H_2(g)$$

The heat generated in this reaction helps melt away grease and the dihydrogen gas released stirs up the solids clogging the drain. Calculate the volume of H_2 formed at 20 °C and 750 torr if 3.12 g of Al is treated with excess NaOH.

3. A certain gaseous hydrocarbon is found to be 88.8% C and 11.2% H by mass. The compound has a density of 2.12 g/L at 31 °C and 742 torr. a) What is the empirical formula of the compound? b) What is the molecular weight of the compound? c) What is the molecular formula of the compound? d) Draw a possible structural formula for the compound.

.888 C .112 H

$\frac{88.8}{12.01}$ $\frac{11.2}{1.01}$

7.4 $\frac{11.1}{7.4} = 1.5$

$C_1 H_{1.5}$

a) C_2H_3

b) $d = \frac{MW\,P}{RT}$ 2.12 = MW(.976) / .08206 · 304.15 = 54.2 g

c C_4H_6

d)

H₂C=CH–CH=CH₂

ChemActivity 34

Enthalpy of Atom Combination

Model 1: Nuclei are Held Together by Coulombic Attraction to Electrons.

Consider two *bare* nuclei, in this case two protons, as shown in Figure 1. From Coulomb's law we know that these protons will repel each other.

Figure 1. Two bare nuclei repel each other.

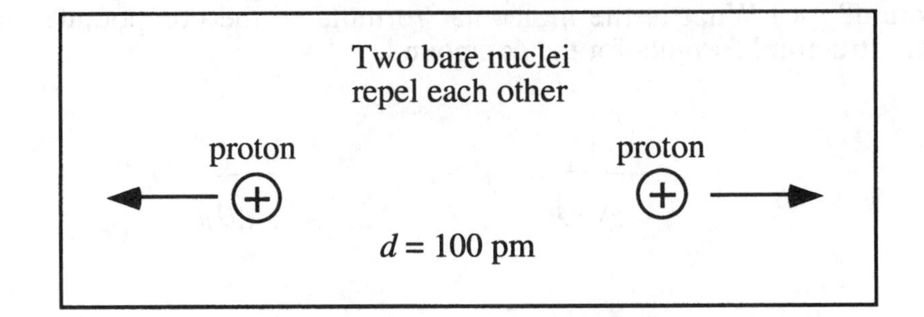

The (repulsive) Coulombic potential energy is (see **ChemActivity 3**):

$$V = \frac{k\, q_1\, q_2}{d} = \frac{2.31 \times 10^{-16}\ \text{J}\bullet\text{pm}}{100\ \text{pm}} = 2.31 \times 10^{-18}\ \text{J} \quad \text{(repulsive)} \tag{1}$$

If an electron is placed between the nuclei, then each nucleus is attracted toward the electron.

Figure 2. Nuclei are attracted to an electron.

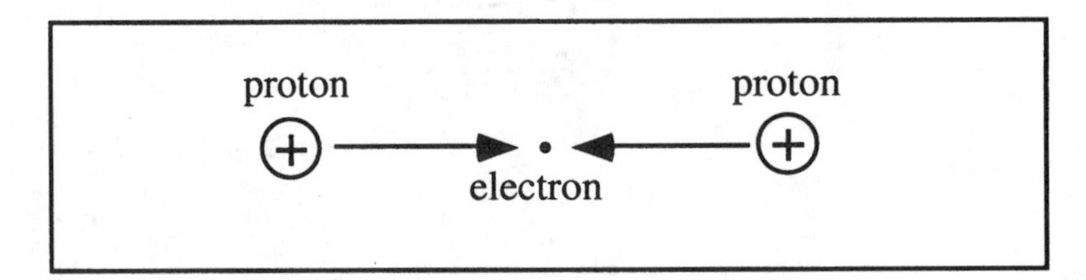

In this simple model, the (attractive) Coulombic potential energy is

$$V = \frac{2 \times (-2.31 \times 10^{-16}\ \text{J}\bullet\text{pm})}{50\ \text{pm}} = -9.24 \times 10^{-18}\ \text{J} \quad \text{(attractive)} \tag{2}$$

The net Coulombic potential energy is

$$V = 2.31 \times 10^{-18}\ \text{J} + (-9.24 \times 10^{-18}\ \text{J}) = -6.93 \times 10^{-18}\ \text{J} \quad \text{(attractive)}$$

On balance, then, the nuclei are held together by the sharing of one or more (usually some multiple of two) electrons.

Critical Thinking Questions

1. Why is "100 pm" in the denominator of equation 1?

2. Why is there a "2" in the numerator of equation 2, but not in equation 1?

3. Why is the numerator in equation 1 positive, but the numerator in equation 2 is negative?

4. The net attractive potential energy in Figure 2 is -6.93×10^{-18} J. Will energy be required to separate the nuclei or will energy be released upon separation? Explain.

Information

This model, of course, only approximates reality. One cannot simply place a stationary electron between two nuclei. Electrons move (have kinetic energy) and occupy certain regions of space (domains or orbitals). Nonetheless, the model above demonstrates that nuclei can be held together by electron sharing between nuclei.

Model 2: Endothermic and Exothermic Processes.

When chemical processes occur, energy (typically in the form of heat) is either released—an **exothermic** process, or absorbed—an **endothermic** process. The breaking of bonds requires energy to pull the atoms apart; bond-breaking is thus an endothermic process. When bonds are formed, energy is released — precisely the same amount of energy which would be required to break those bonds. Thus, the making of bonds is an exothermic process. In most chemical reactions, bonds are both broken and made. Whether the overall reaction is endothermic or exothermic depends on the energy required to perform the needed changes in bonding.

The quantity of energy released or absorbed in a chemical process can be designated by an enthalpy (energy) change, ΔH, for that process. If there is a release of energy when the reaction occurs, the value of ΔH is negative, and the reaction is exothermic. If the reaction results in a net consumption of energy, then ΔH is positive, and the reaction is endothermic.

Figure 1: A simple chemical process.

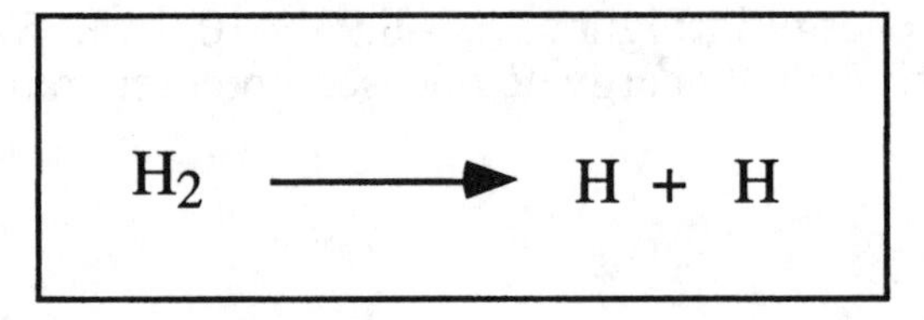

Critical Thinking Questions

5. Is the chemical reaction in Figure 1 exothermic or endothermic?

   endo

6. Is the ΔH for the chemical reaction in Figure 1 positive or negative?

   pos

7. Provide a chemical reaction with a value of ΔH that has the same magnitude of ΔH as the reaction in Figure 1, but has the opposite sign.

   $H + H \rightarrow H_2$

Exercises

1. Predict whether each of the following reactions would be exothermic or endothermic:

   a) $CO(g) \rightarrow C(g) + O(g)$
   b) $2H(g) + O(g) \rightarrow H_2O(g)$
   c) $Na^+(g) + Cl^-(g) \rightarrow NaCl(s)$

2. What is the sign for ΔH in each of the reactions in Ex. 1?

Model 3: Enthalpy of Atom Combination.

When a mole of a compound is produced from its constituent atoms in the gas phase at 1 atmosphere pressure and 25°C, energy is released. The standard state heat (or enthalpy) of atom combination, ΔH°_{ac}, is the difference in enthalpy of product and reactants ($\Delta H^\circ_{product} - \Delta H^\circ_{reactants}$) when this occurs.

Table 1. Standard state enthalpies of atom combination, ΔH°_{ac}.

Substance	ΔH°_{ac} (kJ/mol)	Substance	ΔH°_{ac} (kJ/mol)
H(g)	0	CH_4(g)	–1662.09
C(g)	0		
N(g)	0	H_2O(g)	–926.29
O(g)	0	H_2O(l)	–970.30
H_2(g)	–435.30	NH_3(g)	–1171.76
N_2(g)	–945.408	NO_2(g)	–937.86
O_2(g)	–498.340	N_2O_4(g)	–1932.93

For example, under these conditions, the value of ΔH° for the (hypothetical) reaction

$$C(g) + 4H(g) \rightarrow CH_4(g)$$

is the standard state enthalpy of atom combination for CH_4(g). This is given in Table 1 as –1662.09 kJ/mole. That is, ΔH° = –1662.09 kJ/mole for this reaction. Figure 3 shows, schematically, the enthalpy change for this reaction.

Figure 3. The enthalpy of atom combination of CH_4(g) at 25°C.

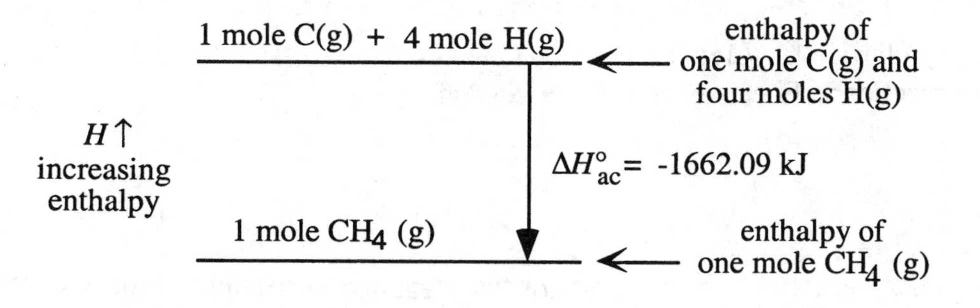

Similarly, the value of ΔH° for the reaction

$$N(g) + 2\,O(g) \rightarrow NO_2(g)$$

can be found in Table 1 as –937.86 kJ/mole.

One can also imagine the process in which a mole of a substance is broken apart into its constituent gas phase atoms. This is precisely the reverse of an "enthalpy of atom combination reaction," and, in this case, energy will be consumed. For example, the value of $\Delta H°$ for the reaction

$$CH_4(g) \rightarrow C(g) + 4H(g)$$

is 1662.09 kJ/mole, as shown in Figure 4.

Figure 4. Breaking one mole of $CH_4(g)$ into its constituent atoms requires energy.

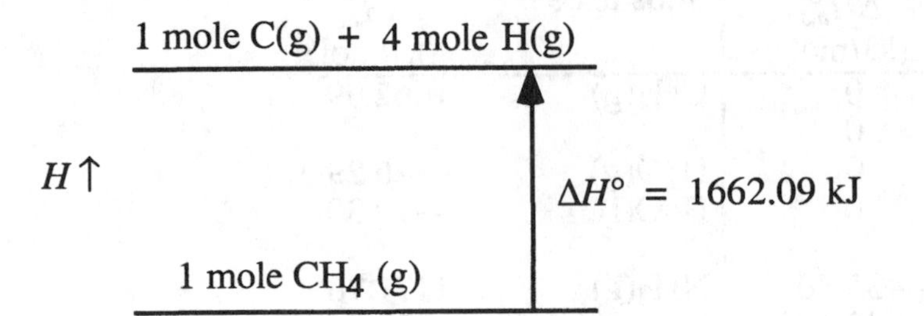

Critical Thinking Questions

8. Why is ΔH°_{ac} of C(g) = 0? Why is ΔH°_{ac} of H(g) = 0?

   do need to break any bonds or form just one atom

9. For molecules, why are all of the values for enthalpies of atom combination negative?

   because forming bonds releases energy making ΔH negative

10. a) Write the Lewis structures for N_2 and O_2.

    b) Explain how these Lewis structures are consistent with the relative enthalpies of atom combination for $N_2(g)$ and $O_2(g)$.

11. What is the value of $\Delta H°$ for the overall process of separating one mole of CH_4 into its constituent atoms, and then reforming one mole of CH_4?

12. a) Calculate the amount of energy released when 2 moles of CH_4 are formed from the appropriate constituent atoms (as opposed to one mole).

b) Based on your answer to part a), does the amount of energy released when a collection of gas phase atoms combines to form molecules depend on the amount of substance present?

Exercises

3. Which of the following heats of atom combination is obviously incorrect?
   a) $CHCl_3(g)$ $\Delta H^\circ_{ac} = -1433.84$ kJ/mole
   b) $Cr(g)$ $\Delta H^\circ_{ac} = 0$
   c) $I_2(s)$ $\Delta H^\circ_{ac} = 213.68$ kJ/mole
   d) none of these is obviously incorrect

4. The ΔH°_{ac} of C(graphite) is –716.682kJ/mole and the ΔH°_{ac} of C(diamond) is –714.787kJ/mole. On average, are the bonds stronger in diamond or graphite?

5. Indicate whether the following statement is true or false and explain your reasoning.

   The bonds in $SiCl_4(g)$ are stronger than the bonds in $SnCl_4(g)$.

6. Use the table below.

Substance	ΔH°_{ac} (kJ/mole)
$H_2O(g)$	–926.29
$H_2S(g)$	–734.74

   a) Determine the O-H bond energy in H_2O and the S–H bond energy in H_2S.
   b) Based on your answer above, which is the stronger bond, O-H or S-H?
   c) Give a rationale for the relative bond strengths of O-H and S-H found above.

ChemActivity 35

Enthalpy Changes in Chemical Reactions

(Is Energy Released or Consumed When a Chemical Reaction Occurs?)

Model 1: The Enthalpy Change for a Chemical Reaction.

Table 1. Standard state enthalpies of atom combination, ΔH°_{ac}.

Substance	ΔH°_{ac} (kJ/mol)	Substance	ΔH°_{ac} (kJ/mol)
H(g)	0		
N(g)	0		
O(g)	0		
$H_2(g)$	–435.30	$NH_3(g)$	–1171.76
$N_2(g)$	–945.408	$NO_2(g)$	–937.86
$O_2(g)$	–498.340	$N_2O_4(g)$	–1932.93

To determine the overall value of ΔH° for a chemical reaction, one can consider the reaction to take place by breaking apart all of the reactant molecules into their constituent atoms, and then reassembling those atoms into the product molecules. Although (in general) this is not the actual process that takes place when chemical reactions occur, thinking about the reaction in this manner is a valid way to determine the value of ΔH° for the reaction.

Figure 1. The enthalpy diagram for the chemical reaction:

$$N_2O_4(g) \rightarrow 2\,NO_2(g)$$

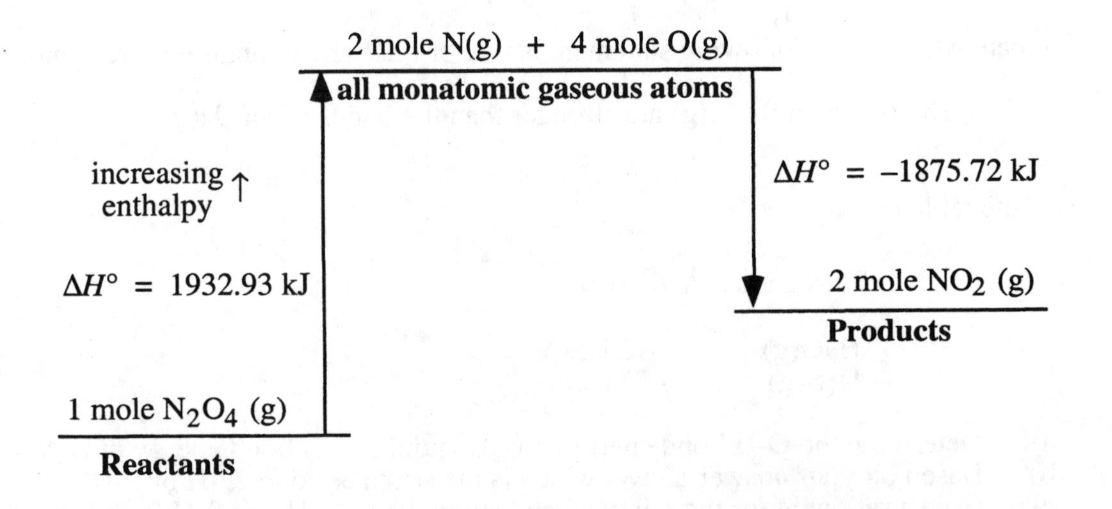

Critical Thinking Questions

1. How much energy is required to break one mole of $N_2O_4(g)$, the reactants, into gaseous atoms?

   1932.93

2. Why is the $\Delta H°$ associated with the upward arrow (left-side of Model 1) a positive number?

   because your breaking bonds

3. How much energy is released when two moles of $NO_2(g)$, the products, are formed from gaseous atoms?

   -1875.72

4. Why is the $\Delta H°$ associated with the downward arrow (Model 1) a negative number?

   because forming bonds

5. For the overall reaction:

   a) is energy released or required?

   ~~released~~ required

   b) is the reaction endothermic or exothermic?

   end

6. Based on the information in Figure 1, what is $\Delta H°$ for the following reaction?

   $N_2O_4(g) \rightarrow 2\ NO_2(g)$

   57.21 kJ

7. For the reaction:

   $$N_2(g) + 3\ H_2(g) \rightarrow 2\ NH_3(g)\ ,$$

   a) make a diagram similar to that in Figure 1.

   2 moles N + 6 moles H

   2251.308

   ΔH = ~~1350.71~~

   1 mol N₂(g) + 3H₂(g) moles

   reactants

   -2343.52

   ΔH ~~1171.76~~

   2 moles NH₃(g)

b) calculate ΔH° based on your diagram.

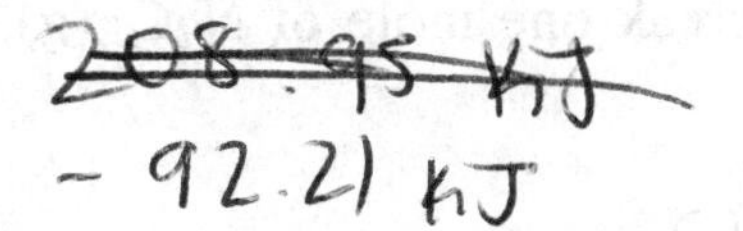

8. Using grammatically correct sentences, describe how to calculate the $\Delta H°$ for the reaction in CTQ 7 given the ΔH°_{ac} of the three species.

9. For any given chemical reaction, if the sum of the enthalpies of atom combination for all of the reactants is more negative than the sum of the enthalpies of atom combination for all of the products, will the value of $\Delta H°$ for the reaction be positive or negative? Explain your reasoning.

Exercises

1. Calculate $\Delta H°$ for each of the following reactions:
   a) $MgO(s) + H_2O(l) \rightarrow Mg(OH)_2(s)$
   b) $2\ Zn(s) + O_2(g) \rightarrow 2\ ZnO(s)$
   c) $TiCl_4(g) + 2\ H_2O(g) \rightarrow TiO_2(s) + 4\ HCl(g)$
2. J. N. Spencer, G. M. Bodner, and L. H. Rickard, *Chemistry: Structure & Dynamics*, Third Edition, John Wiley & Sons, 2006. Chapter 7: Problems: 18, 22, 25, 26, 57, 59, 61, 63, 67, 71, 101, 105.

Model 2: Bond Strength and Enthalpies of Atom Combination.

Recall that for bonds between pairs of atoms, "the stronger the bond, the shorter the bond length." That is, a C–O double bond is stronger than a C–O single bond, and the double bond is also shorter. For bonds between similar atoms, we also find that "the shorter the bond length, the stronger the bond."

Critical Thinking Questions

10. Consider H–F, H–Cl, and H–Br.

    a) Based on bond lengths, which do you expect to have the strongest bond?

    b) Which has the weakest bond? Are your predictions consistent with the ΔH°_{ac} data for HF(g), HCl(g), HBr(g) [see Table A.3 in the Appendix]? Explain your reasoning.

    c) Examine the ΔH°_{ac} data for these species [see Table A.3 in the Appendix] and explain how your answers to parts a) and b) are (or are not) consistent with these values.

11. a) Based on bond lengths, which do you expect to be the stronger bond, C–H or C–Cl?

    b) Examine the ΔH°_{ac} data for CH_4(g) and $CH_3Cl_{(g)}$ and explain how your answer to part a) is (or is not) consistent with these values.

    c) Based on the ΔH°_{ac} values for CH_4(g) and $CH_3Cl_{(g)}$, predict ΔH°_{ac} for CH_3F(g) and CH_3Br(g). Explain your reasoning.

Exercises

3. For each of the following groups of compounds, give the Lewis structures and predict which molecule will have the most negative ΔH°_{ac}: a) Cl_2, Br_2, I_2. b) N_2, P_2, As_2. Explain.

4. Which do you predict has the stronger bond, C–H or C–Cl? Calculate the average C–H bond energy in CH_4 from ΔH°_{ac}. Calculate the average C–Cl bond energy in CCl_4 from ΔH°_{ac}. Compare the two bond energies. Is this the result you predicted?

5. The O–H bond energy in H_2O is 464 kJ/mole. Do you expect the C–H bond energy in CH_4 to be less than or greater than the O–H bond energy? Explain. Is your prediction consistent with the ΔH°_{ac} data? Explain your reasoning.

6. J. N. Spencer, G. M. Bodner, and L. H. Rickard, *Chemistry: Structure & Dynamics*, Third Edition, John Wiley & Sons, 2006. Chapter 7: Problems: 103, 104.

Problem

1. As mentioned previously, molecules attract each other. The forces of attraction between molecules are called intermolecular forces. Consider the following transformations:

$$CH_3OH(l) \rightarrow CH_3OH(g)$$
$$H_2O(l) \rightarrow H_2O(g)$$
$$SO_3(s) \rightarrow SO_3(l)$$
$$SO_3(s) \rightarrow SO_3(g)$$

Calculate the value of ΔH° for each of these reactions. Based on these results, in which phase (gas, liquid, solid) are the intermolecular forces the weakest? The strongest? Explain your reasoning.

ChemActivity 36

Rates of Chemical Reactions (I)

(What Is the Rate of a Chemical Reaction?)

Information

The rate of a chemical reaction depends on how quickly reactants are consumed or, alternatively, how quickly products are formed. By convention, rates of reaction, rates of consumption, and rates of production are always reported as positive numbers.

$$\text{rate of consumption of reactant} = -\frac{\text{change in molarity of a reactant}}{\text{change in time}} = -\frac{\Delta(\text{reactant})}{\Delta\text{time}} \qquad (1)$$

Critical Thinking Questions

1. If time is measured in seconds, what are the units for a rate of consumption?

2. What do the symbols Δ and ( ) in equation (1) represent?

3. Why is there a negative sign in equation (1)?

4. Provide an expression analogous to equation (1) for the rate of production of a product.

Model: The Rate of a Chemical Reaction.

$$3\ ClO^-(aq) \rightarrow 2\ Cl^-(aq) + ClO_3^-(aq) \tag{2}$$

The reaction described in equation (2) was carried out in an aqueous solution with a volume of 2.00 liters. Table 1 displays some data relating to that experiment.

Table 1. Experimental data for equation (2) in a 2.00 liter flask.

Time (s)	Moles of ClO^-	Moles of Cl^-	Moles of ClO_3^-
0	2.40	0	0
1.00×10^2	1.80		

Critical Thinking Questions

5. Fill in the missing entries in Table 1.

6. Based on the data in Table 1, calculate the following:

   a) rate of consumption of ClO^-

   b) rate of production of Cl^-

   c) rate of production of ClO_3^-

Information

The rate of a reaction is defined to be the rate of consumption of a reactant (or the rate of production of a product) whose stoichiometric coefficient is 1 in the balanced chemical equation describing the process. The rate of a reaction can be expressed in terms of the rate of change of concentration of any of the reactants or products.

Critical Thinking Questions

7. Based on the answers to CTQ 6, what is the rate of the reaction in the model?

8. What is the ratio of:

   a) the rate of consumption of ClO^- to the rate of reaction?

   b) the rate of production of Cl^- to the rate of reaction?

   c) the rate of production of ClO_3^- to the rate of reaction?

9. In general, how is the rate of reaction related to the rate of change of concentration of a reactant or product and the corresponding stoichiometric coefficient? Provide your answer using a grammatically correct English sentence and in the form of an equation.

Exercises

1. Indicate whether the following statement is true or false and explain your reasoning:

   a) The rate of a reaction is equal to the rate at which each of the products is formed.

   b) If PCl_5 decomposes according to the reaction $PCl_5(g) \rightarrow PCl_3(g) + Cl_2(g)$, the rate of consumption of PCl_5 is twice the rate of production of Cl_2.

2. For the reaction $N_2(g) + 3\ H_2(g) \rightarrow 2\ NH_3(g)$, the rate of consumption of H_2 was observed to be 3.50×10^{-4} M/s under certain conditions. Determine the rate of production of ammonia. Determine the rate of this reaction.

3. For the reaction $2\ O_3(g) \rightarrow 3\ O_2(g)$, the rate of production of O_2 was observed to be 1.35×10^{-4} M/s under certain conditions. Determine the rate of consumption of ozone and the rate of this reaction.

4. For the reaction $3\ I^-(aq) + IO_2^-(aq) + 4\ H^+(aq) \rightarrow 2\ I_2(aq) + 2\ H_2O$, the rate of production of H_2O was observed to be 5.0×10^{-2} M/s under certain conditions. Determine the rate of consumption of I^-, IO_2^-, and H^+. Determine the rate of production of I_2. What is the rate of this reaction?

5. J. N. Spencer, G. M. Bodner, and L. H. Rickard, *Chemistry: Structure & Dynamics*, Third Edition, John Wiley & Sons, 2006. Chapter 10: Problem: 9.

6. J. N. Spencer, G. M. Bodner, and L. H. Rickard, *Chemistry: Structure & Dynamics*, Third Edition, John Wiley & Sons, 2006. Chapter 14: Problems: 18-20, 22-24.

ChemActivity 37

Equilibrium (I)

(Do Reactions Ever Really Stop?)

Model 1: The Conversion of cis-2-butene to trans-2-butene.

Consider a simple chemical reaction where the forward reaction occurs in a single step and the reverse reaction occurs in a single step:

$$A \rightleftarrows B$$

The following chemical reaction, where cis-2-butene is converted into trans-2-butene, is an example.

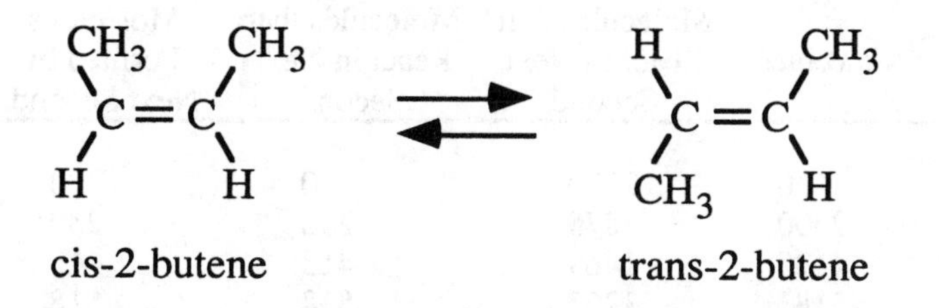

In this example, one end of a cis-2-butene molecule rotates 180° to form a trans-2-butene molecule. Rotation around a double bond rarely happens at room temperature because the collisions are not sufficiently energetic to weaken the double bond. At higher temperatures, around 400°C for cis-2-butene, collisions are sufficiently energetic and an appreciable reaction rate is detected.

Critical Thinking Questions

1. Make a model of cis-2-butene with a modeling kit. What must be done to convert cis-2-butene to trans-2-butene?

2. Make a model of trans-2-butene with a modeling kit. What must be done to convert trans-2-butene to cis-2-butene?

3. A large number of cis-2-butene molecules is placed in a container.

   a) Predict what will happen (that is, what the final composition will be) if these molecules are allowed to stand at room temperature for a long time.

   b) Predict what will happen if these molecules are allowed to stand at 400 °C for a long time.

Model 2: The Number of Molecules as a Function of Time.

Consider the simple reaction:

$$A \rightleftarrows B$$

The system is said to be at *equilibrium* when the concentrations of reactants and products stops changing.

Imagine the following hypothetical system. Exactly 10,000 A molecules are placed in a container which is maintained at 800°C. We have the ability to monitor the number of A molecules and the number of B molecules in the container at all times. We collect the data at various times and compile Table 1.

Table 1. Number of A and B molecules as a function of time.

Time (seconds)	Number of A Molecules	Number of B Molecules	Number of A Molecules that React in Next Second	Number of B Molecules that React in Next Second	Number of A Molecules Formed in Next Second	Number of B Molecules Formed in Next Second
0	10000	0	2500	0	0	2500
1	7500	2500	1875	250	250	1875
2	5875	4125	1469	413	413	1469
3	4819	5181	1205	518	518	1205
4	4132	5868	1033	587	587	1033
5	3686	6314	921	631	631	921
6	3396	6604	849	660	660	849
7	3207	6793	802	679	679	802
8	3085	6915	771	692	692	771
9	3005	6995	751	699	699	751
10	2953	7047	738	705	705	738
11	2920	7080	730	708	708	730
12	2898	7102	724	710	710	724
13	2884	7116	721	712	712	721
14	2874	7126	719	713	713	719
15	2868	7132	717	713	713	717
16	2864	7136	716	714	714	716
17	2862	7138	715	714	714	715
18	2860	7140	715	714	714	715
19	2859	7141	715	714	714	715
20	2858	7142	715	714	714	715
21	2858	7142	714	714	714	714
22	2858	7142	714	714	714	714
23	2857	7143	714	714	714	714
24	2857	7143	714	714	714	714
25	2857	7143	714	714	714	714
30	2857	7143	714	714	714	714
40	2857	7143	714	714	714	714
50	2857	7143	714	714	714	714

Critical Thinking Questions

4. During the time interval 0 – 1 s:

   a) How many A molecules react?

   b) How many B molecules are formed?

   c) Why are these two numbers equal?

5. During the time interval 10 – 11 s:

   a) How many B molecules react?

   b) How many A molecules are formed?

   c) Why are these two numbers equal?

6. a) During the time interval 0 – 1 s, what fraction of the A molecules react?

   b) During the time interval 10 – 11 s, what fraction of the A molecules react?

   c) During the time interval 24 – 25 s, what fraction of the A molecules react?

   d) During the time interval 40 – 41 s, what fraction of the A molecules react?

7. Based on the answer to CTQ 6, verify that 921 molecules of A react during the time interval 5 – 6 s.

8. During the time interval 100 – 101 s, how many molecules of A react? Explain your reasoning.

9. a) During the time interval 1 – 2 s, what fraction of the B molecules react?

   b) During the time interval 10 – 11 s, what fraction of the B molecules react?

   c) During the time interval 24 – 25 s, what fraction of the B molecules react?

   d) During the time interval 40 – 41 s, what fraction of the B molecules react?

10. Based on the answer to CTQ 9, verify that 631 molecules of B react during the time interval 5 – 6 s.

11. During the time interval 100 – 101 s, how many molecules of B react? Explain your reasoning.

12. For the reaction described in Table 1:

    a) How long did it take for the reaction to come to equilibrium?

    b) Are A molecules still reacting to form B molecules at t = 500 seconds?

    c) Are B molecules still reacting to form A molecules at t = 500 seconds?

Information

For the process in Model 2,

$$\text{rate of conversion of B to A} = \text{number of B molecules that react per second} = \frac{\Delta \text{ number of B molecules}}{\Delta t}$$

The relationship between the rate of conversion of B to A and the number of B molecules is given by equation (1):

$$\text{rate of conversion of B to A} = k_B \times \text{number of B molecules} \qquad (1)$$

where k_B is a specific value.

13. What is the value of k_B in equation (1)? Be sure to include units in your answer.

14. a) Write a mathematical equation (analogous to equation (1)) that relates the rate of conversion of A molecules into B molecules to the number of A molecules present. This equation should include a constant k_A.

    b) What is the value of k_A (include units)?

Exercises

1. Describe, using grammatically correct English sentences, what is meant by the phrase "at equilibrium" as it refers to the chemical process:

   $A \rightleftarrows B$

2. The chemical system $2A \rightleftarrows B$ is at equilibrium. In the next second 344 molecules of A will react to form B molecules. a) How many B molecules will be produced in the next second? b) How many B molecules will react in the next second? c) How many A molecules will be produced in the next second?

ChemActivity 38

Equilibrium (II)

Information

In **ChemActivity 37: Equilibrium (I)**, we saw that

$$\text{rate of conversion of B to A} = \frac{0.10}{\text{s}} \times \text{number of B molecules}$$

and that

$$\text{rate of conversion of A to B} = \frac{0.25}{\text{s}} \times \text{number of A molecules.} \qquad (1)$$

We can easily change equation (1), which has units of molecules/s, to the units $\frac{\text{moles}}{\text{L s}}$

$$\text{number of A molecules} \times \frac{\text{mol}}{6.022 \times 10^{23}\ \text{molecules}} \times \frac{1}{\text{V(in Liters)}} = (\text{A})$$

Thus, equation (1) can be rewritten as

$$\text{rate of conversion of A to B} = \frac{0.25}{\text{s}}(\text{A}) = k_A\,(\text{A}) \qquad (2)$$

We can now write for the previous reaction of A to B:

$$\text{A} \rightarrow \text{B} \qquad \text{rate} = k_A\,(\text{A}) = \frac{0.25}{\text{s}}(\text{A})$$

$$\text{B} \rightarrow \text{A} \qquad \text{rate} = k_B\,(\text{B}) = \frac{0.10}{\text{s}}(\text{B})$$

or

$$\text{A} \rightleftarrows \text{B}$$

$$\text{rate}_{\text{forward}} = k_A\,(\text{A}) = \frac{0.25}{\text{s}}(\text{A})$$

$$\text{rate}_{\text{reverse}} = k_B\,(\text{B}) = \frac{0.10}{\text{s}}(\text{B})$$

where k_A and k_B are called the *specific rate constants* for the forward and reverse reactions.

Model: The Concentration of Molecules as a Function of Time.

Consider the reaction A $\rightleftarrows$ B in **ChemActivity 37: Equilibrium (I)**. If the volume of the container is 1.661×10^{-19} L, we can calculate the concentrations of A and B as a function of time.

Table 1. The concentrations of A and B as a function of time.

Volume of container = 1.661×10^{-19} L

$k_A = 0.25\ s^{-1}$ $k_B = 0.10\ s^{-1}$

Time (seconds)	Number of A Molecules	Number of B Molecules	(A) (M)	(B) (M)	Forward Rate (10^{-2} M s^{-1})	Reverse Rate (10^{-2} M s^{-1})
0	10000	0	0.1000	0.0000	2.50	0.00
1	7500	2500	0.0750	0.0250	1.88	0.25
2	5875	4125	0.0588	0.0413	1.47	0.41
3	4819	5181	0.0482	0.0518	1.20	0.52
4	4132	5868	0.0413	0.0587	1.03	0.59
5	3686	6314	0.0369	0.0631	0.92	0.63
6	3396	6604	0.0340	0.0660	0.85	0.66
7	3207	6793	0.0321	0.0679	0.80	0.68
8	3085	6915	0.0308	0.0692	0.77	0.69
9	3005	6995	0.0301	0.0699	0.75	0.70
10	2953	7047	0.0295	0.0705	0.74	0.70
11	2920	7080	0.0292	0.0708	0.73	0.71
12	2898	7102	0.0290	0.0710	0.72	0.71
13	2884	7116	0.0288	0.0712	0.72	0.71
14	2874	7126	0.0287	0.0713	0.72	0.71
15	2868	7132	0.0287	0.0713	0.72	0.71
16	2864	7136	0.0286	0.0714	0.72	0.71
17	2862	7138	0.0286	0.0714	0.72	0.71
18	2860	7140	0.0286	0.0714	0.72	0.71
19	2859	7141	0.0286	0.0714	0.71	0.71
20	2858	7142	0.0286	0.0714	0.71	0.71
25	2857	7143	0.0286	0.0714	0.71	0.71
30	2857	7143	0.0286	0.0714	0.71	0.71
40	2857	7143	0.0286	0.0714	0.71	0.71
50	2857	7143	0.0286	0.0714	0.71	0.71

Critical Thinking Questions

1. Show that the concentration of A at $t = 5$ s is correct (given that the number of A molecules is 3686).

2. Show that the value given in the "Forward Rate" column is correct (in Table 1) when:

   a) $(A) = (A)_o$ (the initial concentration of A; $t = 0$ s)

   b) $(A) = (A)_5$ (at $t = 5$ s)

   c) $(A) = (A)_e = [A]$ (the equilibrium concentration of A)

3. Show that the value given in the "Reverse Rate" column is correct (in Table 1) when:

   a) $(B) = (B)_o$ (the initial concentration of B; $t = 0$ s)

   b) $(B) = (B)_5$ (at $t = 5$ s)

   c) $(B) = (B)_e = [B]$ (the equilibrium concentration of B)

4. Use a grammatically correct English sentence to describe the realtionship between the forward rate and reverse rate at equilibrium.

Information

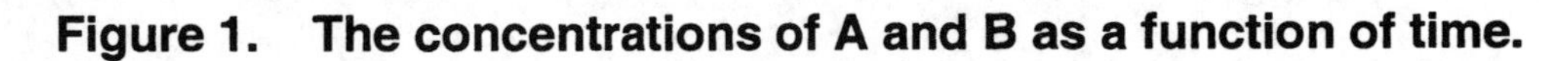

Figure 1. The concentrations of A and B as a function of time.

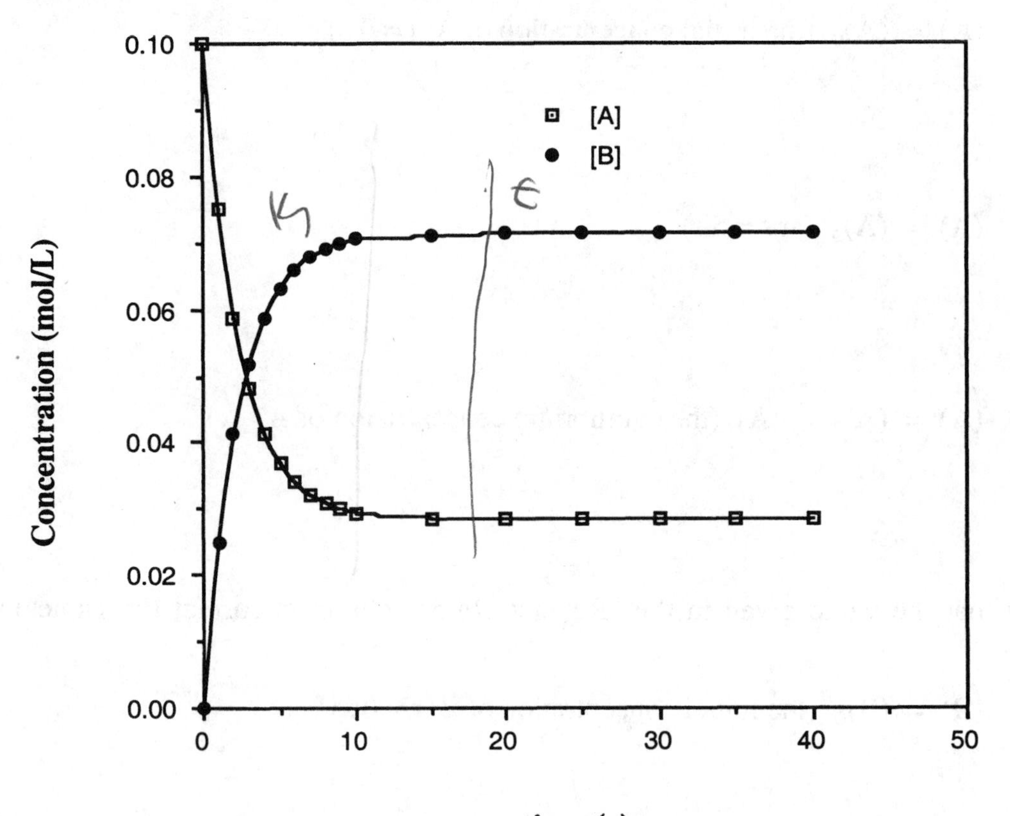

Critical Thinking Questions

5. The data in Figure 1 can be divided into two regions—the kinetic region and the equilibrium region.

   a) Indicate these regions (time intervals) on the figure.

   b) Use a grammatically correct English sentence to describe the kinetic region.

6. From the graph <u>only</u>, estimate the value of $\frac{(B)}{(A)}$ at:

   a) $t = 1$ s

   b) $t = 4$ s

   c) $t = 15$ s

   d) $t = 20$ s

   e) $t = 40$ s

7. In what region (kinetic or equilibrium) is the quantity $\frac{(B)}{(A)}$ a constant?

Exercises

1. Use Table 1 to calculate the value of $\frac{(B)}{(A)}$ at:

   a) $t = 1$ s

   b) $t = 4$ s

   c) $t = 15$ s

   d) $t = 20$ s

   e) $t = 40$ s

   Why is there a small difference between these values and your values in CTQ 6?

2. Suppose that the container and starting concentrations of A and B are identical to those in Table 1, but the values of the rate constants were changed to $k_A = 0.10\ s^{-1}$ and $k_B = 0.25\ s^{-1}$. What would be the equilibrium concentrations [A] and [B]?

3. J. N. Spencer, G. M. Bodner, and L. H. Rickard, *Chemistry: Structure & Dynamics*, Third Edition, John Wiley & Sons, 2006. Chapter 10: Problems: 3, 31.

Problem

1. Examine the following graph, which describes a chemical reaction involving A, B, and C.

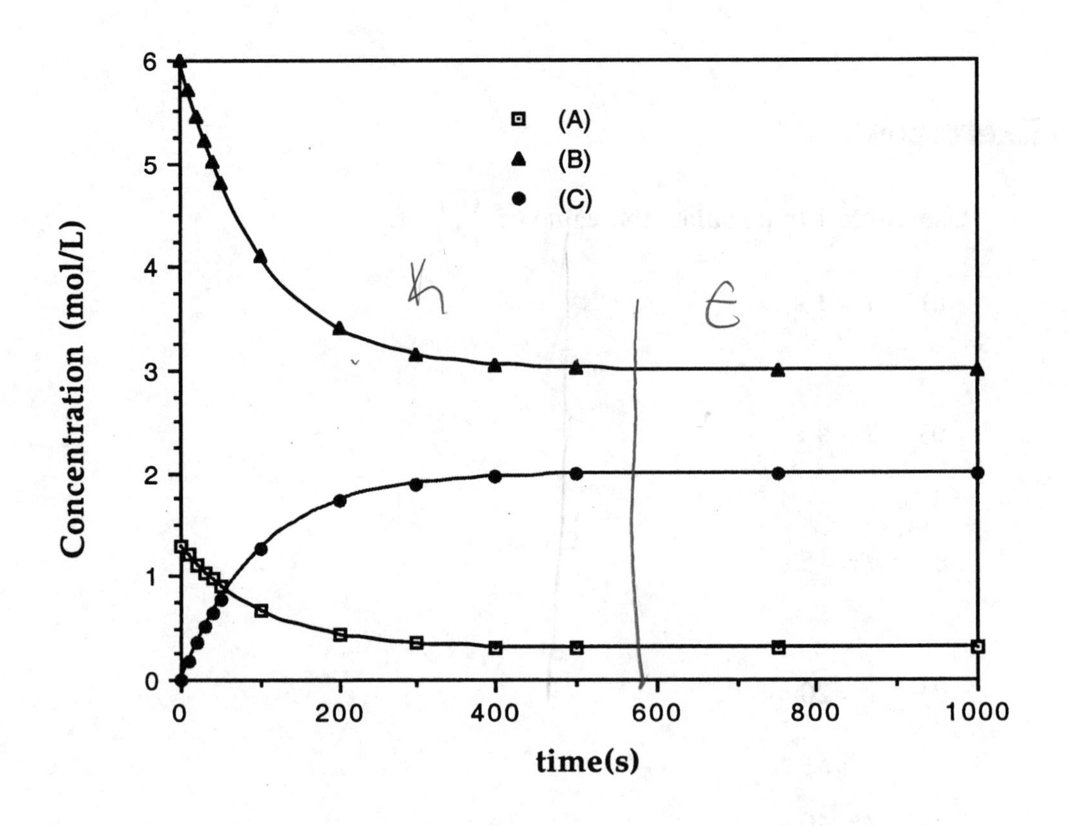

a) Indicate the kinetic region and the equilibrium region on the graph.
b) Estimate $(A)_o$, $(B)_o$, $(C)_o$.
c) Estimate [A], [B], [C].
d) Write a balanced chemical equation for this reaction.

ChemActivity 39

The Equilibrium Constant (I)

(Why Do Some Things Never Change?)

Model 1: A Simple Gas Phase Reaction, Y(g) ⇄ Z(g).

Consider a simple reaction:

$$Y(g) \rightleftarrows Z(g)$$

$$\text{rate}_{\text{forward}} = k_Y \,(Y) \qquad (1)$$

$$\text{rate}_{\text{reverse}} = k_Z \,(Z) \qquad (2)$$

The system is said to be at equilibrium when the concentrations of reactants and products stop changing. At equilibrium,

$$(Y) = (Y)_e = [Y] \quad \text{and} \quad (Z) = (Z)_e = [Z]$$

Table 1. Results of various experiments on the system Y ⇄ Z at some temperature. Molecules "Y" and "Z" are different for each set.

Set	Trial	$(Y)_o$	$(Z)_o$	k_Y	k_Z	[Y]	[Z]
M	1	100	0	0.20	0.40	67	33
	2	50	50	0.20	0.40	67	33
N	1	200	0	0.60	0.20	50	150
	3	50	150	0.60	0.20	50	150
O	1	100	0	0.50	0.50	50	50
P	1	80	20	0.20	0.60	75	25
	2	59	41	0.20	0.60	75	25

Critical Thinking Questions

1. What is the distinction between $(Y)_o$ and $[Y]$ in the column heading of Table 1?

2. For Trial M2, calculate the rate of the forward reaction ($Y \rightarrow Z$) at equilibrium and the rate of the reverse reaction ($Z \rightarrow Y$) at equilibrium. How do these values compare?

3. For Trial P2, calculate the rate of the forward reaction ($Y \rightarrow Z$) at equilibrium and the rate of the reverse reaction ($Z \rightarrow Y$) at equilibrium. How do these values compare?

4. In general, how does the rate of the forward reaction ($Y \rightarrow Z$) at equilibrium compare to the rate of the reverse reaction ($Z \rightarrow Y$) at equilibrium?

5. Examine Table 1.

   a) If $k_Y < k_Z$, what are the relative values of the equilibrium concentrations of Y and Z? Explain.

   b) If $k_Y > k_Z$, what are the relative values of the equilibrium concentrations of Y and Z? Explain.

6. Use the general result from CTQ 4, along with equations (1) and (2) (where k_Y and k_Z are constants), to show that $\frac{[Z]}{[Y]}$ is a constant.

Model 2: The Law of Mass Action.

The Law of Mass Action states that for a chemical system described by the balanced chemical equation

$$aA + bB \rightleftarrows cC + dD$$

the ratio $\frac{[C]^c [D]^d}{[A]^a [B]^b}$ is a constant at a given temperature. The ratio is called the **equilibrium constant expression**, and the numerical value of the ratio is called the **equilibrium constant**, K_c. Note that a, b, c, and d are the stoichiometric coefficients in the balanced chemical equation.

$$K_c = \frac{[C]^c [D]^d}{[A]^a [B]^b}$$

By convention, equilibrium constant values are given without units.

For example:

the reaction: $2H_2(g) + O_2(g) \rightleftarrows 2H_2O(g)$

the equilibrium expression: $$K_c = \frac{[H_2O]^2}{[H_2]^2[O_2]} \qquad (3)$$

the equilibrium constant (25°C): $K_c = 10^{83}$ (experimental number)

Critical Thinking Questions

7. When a mixture of $H_2(g)$, $O_2(g)$, and $H_2O(g)$ reaches equilibrium, what species is present in the largest amount? Explain your reasoning.

8. Use data set M of Table 1. Write the equilibrium constant expression. Find the value of the equilibrium constant. Find [Y] and [Z] given $(Y)_o = 200$ and $(Z)_o = 0$.

Model 3: Two Related Chemical Reactions.

$$PCl_3(g) + Cl_2(g) \rightleftarrows PCl_5(g) \qquad (4)$$

$$PCl_5(g) \rightleftarrows PCl_3(g) + Cl_2(g) \qquad (5)$$

Critical Thinking Questions

9. a) Provide an expression for the K_c of reaction (1).

   b) Provide an expression for the K_c of reaction (2).

   c) What is the relationship between reactions (1) and (2) ?

   d) What is the relationship between the equilibrium expression for reaction (1) and the expression for reaction (2) ?

Exercises

1. Explain why, if $k_Y > k_Z$, then [Z] > [Y].
2. The equilibrium state is often described as being "dynamic equilibrium." What do you think the word "dynamic" refers to?
3. Calculate the value of K_c for each of the sets of trials in Table 1. For each data set, does K_c depend on the initial concentrations of Y and Z?
4. Write the equilibrium constant expression, K_c, for each of the following reactions:

   a) $2HI(g) \rightleftarrows H_2(g) + I_2(g)$

   b) $3H_2(g) + N_2(g) \rightleftarrows 2NH_3(g)$

c) $\frac{3}{2}H_2(g) + \frac{1}{2}N_2(g) \rightleftarrows NH_3(g)$

d) cis-2-butene(g) $\rightleftarrows$ trans-2-butene(g)

e) $O_3(g) \rightleftarrows O_2(g) + O(g)$

f) $Xe(g) + 2F_2(g) \rightleftarrows XeF_4(g)$

5. Find the mathematical relationship between the equilibrium constant expressions of reaction b) and reaction c) in exercise 4.

6. When the following reaction reaches equilibrium,

$$A(g) + 2B(g) \rightleftarrows C(g)$$

the following concentrations are measured: [A] = 0.60; [B] = 0.20; [C] = 0.55. What is the value of K_c for this reaction?

7. An equilibrium mixture of PCl_5, PCl_3,and Cl_2, at a certain temperature in a 5.0 L container consists of 0.80 mole PCl_5, 0.55 mole PCl_3, and 1.2 mole Cl_2. Calculate K_c for the reaction:

$$PCl_3(g) + Cl_2(g) \rightleftarrows PCl_5(g)$$

8. Calculate K_c for the reaction:

$$3H_2(g) + N_2(g) \rightleftarrows 2NH_3(g)$$

given that the equilibrium concentrations are: $[H_2] = 1.5$; $[NH_3] = 0.24$; $[N_2] = 2.5$.

9. $K_c = 150.0$ at a certain temperature for the reaction:

$$2NO(g) + O_2(g) \rightleftarrows 2NO_2(g)$$

What is the concentration of NO_2 if the equilibrium concentration of NO and O_2 are 1.00×10^{-3} and 5.00×10^{-2}, respectively?

10. The following gases are added to a 1.00 L container: 2.0 mole of A; 4.0 mole B. These gases react as follows:

$$A(g) + 3B(g) \rightleftarrows C(g) + 2D(g)$$

At equilibrium, the container contains 0.4 moles of D.

a) Calculate the moles of A, B, and C in the container at equilibrium.
b) Calculate the concentrations of A, B, C, and D at equilibrium.
c) Calculate the value of the equilibrium constant, K_c, for this reaction.

11. A 1.00 L flask contains an equilibrium mixture of 24.9 g of N_2, 1.35 g of H_2, and 2.15 g of NH_3 at some temperature. Calculate the equilibrium constant for the reaction at this temperature.

$$3\,H_2(g) + N_2(g) \rightleftarrows 2\,NH_3(g)$$

12. The reaction $2NO(g) + O_2(g) \rightleftarrows 2NO_2(g)$ has $K_c = 100.0$.

    What is the value for K_c for the reaction $2NO_2(g) \rightleftarrows 2NO(g) + O_2(g)$?

13. Write the equilibrium constant expressions, K_c, for

$$3\,H_2(g) + N_2(g) \rightleftarrows 2\,NH_3(g)$$

$$\frac{3}{2}H_2(g) + \frac{1}{2}N_2(g) \rightleftarrows NH_3(g)$$

    a) If $K_c = 0.78$ for the first reaction, what is K_c for the second reaction (at the same temperature)?

    b) What is the value of K_c for the reaction (at the same temperature)?

$$NH_3(g) \rightleftarrows \frac{3}{2}H_2(g) + \frac{1}{2}N_2(g)$$

14. The equilibrium constants at some temperature are given for the following reactions:

$$2\,NO(g) = N_2(g) + O_2(g) \qquad K_c = 2.4 \times 10^{-18}$$

$$NO(g) + \frac{1}{2}Br_2(g) = NOBr(g) \qquad K_c = 1.4$$

    Using this information, determine the value of the equilibrium constant for the following reaction at the same temperature:

$$\frac{1}{2}N_2(g) + \frac{1}{2}O_2(g) + \frac{1}{2}Br_2(g) = NOBr(g)$$

15. J. N. Spencer, G. M. Bodner, and L. H. Rickard, *Chemistry: Structure & Dynamics*, Third Edition, John Wiley & Sons, 2006. Chapter 10: Problems: 17-23, 25, 26.

Problems

1. Indicate whether the following statement is true or false and explain your reasoning.

   The value of K_c for the reaction $2\ AB(g) + B_2(g) \leftrightarrows 2\ AB_2(g)$ must be less than the value of K_c for the reaction $2\ AB_2(g) \leftrightarrows 2\ AB(g) + B_2(g)$.

2. Consider the following reaction:

   $$2\ A(g) \rightleftarrows B(g)$$

   One mole of A was placed in a 1.0 L flask and the reaction was followed as a function of time. The data are shown in the figure below:

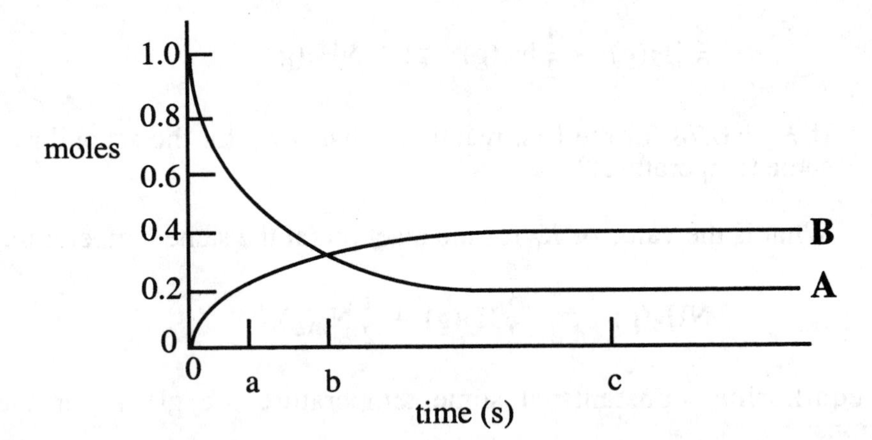

   a) Determine the value of the equilibrium constant, K_c.
   b) Which is larger at time "a", the forward rate or the reverse rate? Briefly explain.
   c) Which is larger at time "b", the forward rate or the reverse rate? Briefly explain.
   d) Which is larger at time "c", the forward rate or the reverse rate? Briefly explain.

ChemActivity 40

The Reaction Quotient

(Are We at Equilibrium Yet?)

Model 1: A Chemical System at Equilibrium.

$$PCl_3(g) + Cl_2(g) \rightleftarrows PCl_5(g)$$

A 5.00-liter box at 25°C has 0.0500 moles of $PCl_3(g)$, 0.0200 moles of $Cl_2(g)$, and 0.200 moles of $PCl_5(g)$. It is known that:

$$K_c = 1.00 \times 10^3 = \frac{[PCl_5]}{[PCl_3]\,[Cl_2]} \qquad (1)$$

Critical Thinking Questions

1. What is the concentration of $PCl_3(g)$ in the system described in Model 1?

2. Verify that the reaction occurring in the box described above is at equilibrium.

Model 2: More Reactant Is Added.

A needle is inserted into the box described in Model 1 and an additional 0.0600 moles of PCl_3 are injected into the reaction mixture.

Critical Thinking Questions

3. At the instant of injection (before any chemical reaction takes place):

   a) What is the total number of moles of PCl_3 in the box?

   b) What is the new concentration of PCl_3 in the box?

   c) Is the system now at equilibrium? Explain.

4. Predict which of the following will happen to the moles of PCl_5 after injection of the 0.0600 moles of PCl_3.

   a) No change in the moles of PCl_5 because the system is at equilibrium.
   b) PCl_3 and Cl_2 will be consumed to form more PCl_5.
   c) PCl_5 will be consumed to form more PCl_3 and Cl_2.

   Explain your reasoning.

Model 3: The Reaction Quotient.

The **reaction quotient**, Q_c, for the reaction, $aA + bB \rightleftarrows cC + dD$ is defined as follows:

$$Q_c = \frac{(C)^c (D)^d}{(A)^a (B)^b}$$

Note that the reaction quotient expression *looks* similar to the equilibrium constant expression. The difference is that the reaction quotient can be calculated at any time during the reaction—at equilibrium or not at equilibrium. For reactions involving liquids or solids, the corresponding reaction quotient, Q, omits those species (the same species that are not included in the equilibrium constant expression).

The reaction quotient for the conditions in Model 2 (after the 0.0600 moles of PCl_3 have been added) is

$$Q_c = \frac{(PCl_5)}{(PCl_3)(Cl_2)} = 4.55 \times 10^2$$

Critical Thinking Questions

5. Verify that $Q_c = 4.55 \times 10^2$ in Model 2.

6. Is the reaction mixture at equilibrium? If not, what will happen?

   a) No change because the system is at equilibrium.
   b) PCl_3 and Cl_2 must be consumed to form more PCl_5.
   c) PCl_5 must be consumed to form more PCl_3 and Cl_2.

   Explain your reasoning.

7. Consider a situation in which the reaction quotient, Q_c, for a given reaction is larger than the equilibrium constant, K_c.

   a) How must the value of Q_c change to reach equilibrium?

   b) Describe how the concentrations of reactants and/or the concentration of products must change to reach equlibrium.

8. Explain why the reaction quotient is useful.

Information

Consider the following reaction:

$$PCl_3(g) + Cl_2(g) \rightleftarrows PCl_5(g)$$

A 5.00-liter box at 25°C has 0.0500 moles of $PCl_3(g)$, 0.0200 moles of $Cl_2(g)$, and 0.200 moles of $PCl_5(g)$. A needle is inserted into the box described above and an additional 0.0600 moles of PCl_3 are injected into the reaction mixture. (This is the same scenario as in Models 1 and 2.)

Critical Thinking Questions

9. At the instant of injection (before any chemical reaction takes place), the total number of moles of PCl_3 in the box is 0.1100 (see CTQ 3).

   a) How many moles of Cl_2 are present (before any chemical reaction takes place)?

   b) How many moles of PCl_5 are present (before any chemical reaction takes place)?

10. In order to reach equilibrium assume that x moles of PCl_3 react.

    a) How many moles of Cl_2 react?

    b) How many moles of PCl_5 are formed?

c) Complete the following table:

	PCl_3	Cl_2	PCl_5
initial moles	0.1100		
change in moles	$-x$		x

11. How many moles of each species are present at equilibrium? Fill in the appropriate expression for each species in the row "equilibrium moles." (The first two rows are identical to the first two rows in CTQ 10c.)

	PCl_3	Cl_2	PCl_5
initial moles	0.1100		
change in moles	$-x$		x
equilibrium moles	$0.1100 - x$		

12. For the reaction in the model, recall that the total volume is 5.00 L and fill in the appropriate concentrations in the row "equilibrium conc." (The first three rows are identical to the first three rows in CTQ 11.)

	PCl_3	Cl_2	PCl_5
initial moles	0.1100		
change in moles	$-x$		x
equilibrium moles	$0.1100 - x$		
equilibrium conc	$\frac{0.1100 - x}{5.00}$		

13. The following two values of x are obtained using equation (1), the entries in the table in CTQ 12, and the quadratic formula: 0.12 and 9.5×10^{-3}. Based on this information calculate the equilibrium concentrations of PCl_3, Cl_2, and PCl_5.

14. For the reaction in the model, write the equilibrium constant expression and enter the values found in CTQ 13. Verify that the appropriate multiplication and division yields the value of the equilibrium constant. This is a method to verify your answer. If you do not get the value of K_c, you made a mistake somewhere!

15. Assume that the system is at equilibrium as determined in CTQ 13. What will happen to the number of moles of PCl_3 present if some Cl_2 gas is suddenly added to the box? Explain your reasoning.

16. Assume that the system is at equilibrium as determined in CTQ 13. What will happen to the number of moles of PCl_3 present if some PCl_5 gas is suddenly added to the box? Explain your reasoning.

Exercises

1. Calculate the value of x (in CTQ 13) using equation (3), the entries in the Table in CTQ 12, and the quadratic formula.

2. For the following reaction in a 5.0 L reaction vessel and at some temperature:

$$CO_2(g) + H_2(g) \rightleftarrows CO(g) + H_2O(g)$$

$K_c = 0.20$ at this temperature.

Complete the following table (use x where appropriate):

	CO_2	H_2	CO	H_2O
initial moles	1.00	2.00	0	0
change in moles				
equilibrium moles				
equilibrium conc				
equilibrium conc value (no "x")				

Verify that your equilibrium concentrations are correct!

3. The following reaction does not proceed at room temperature (the equilibrium constant is exceedingly low, $\approx 10^{-31}$), but NO is produced at higher temperatures (such as found in automobile engines).

$$N_2(g) + O_2(g) \rightleftarrows 2\,NO(g) \quad .$$

Suppose that 5.00 moles of N_2 and 10.00 moles of O_2 are added to a reaction chamber at room temperature. The temperature is increased to 1000°C. If x moles of N_2 react, how many moles of O_2 react? How many moles of NO are formed?

4. Complete the following table for the reaction $N_2(g) + O_2(g) \rightleftarrows 2\,NO(g)$.

	N_2	O_2	NO
initial moles	5.00	10.00	0
change in moles	$-x$		
equilibrium moles	$5.00 - x$		

5. For the following reaction in a 10.0 L reaction vessel and at some temperature:

$$N_2(g) + 2\,H_2(g) \rightleftarrows N_2H_4(g)$$

$K_c = 5.0 \times 10^{-3}$ at this temperature.

Complete the following table:

	N_2	H_2	N_2H_4
initial moles	1.00	1.50	0
change in moles			
equilibrium moles			
equilibrium conc expression			

Which is the correct concentration of N_2H_4 at equilibrium?

a) 0.11 M

b) 1.1×10^{-3} M

c) 1.1×10^{-5} M.

What is the concentration of N_2 at equilibrium? Of H_2?

6. $K_c = 1.60$ at 986°C for the following reaction:

$$CO_2(g) + H_2(g) \rightleftarrows CO(g) + H_2O(g)$$

Complete the following table for a 1.00 L vessel:

	CO_2	H_2	CO	H_2O
initial moles	1.00	2.00	1.00	2.00
change in moles				
equilibrium moles				
equilibrium conc				
equilibrium conc value				

Verify that your equilibrium concentrations are correct!

7. Consider the equilibrium process

$$2NH_3(g) \rightleftarrows N_2(g) + 3H_2(g)$$

An otherwise empty 2.0-liter container is filled with 2.65 moles of $NH_3(g)$ and the system is allowed to come to equilibrium at some temperature. At equilibrium, there are 1.26 moles of $H_2(g)$ present. Complete the table (note that x is not required here, all numerical values can be used).

	NH_3	N_2	H_2
initial moles	2.65		
change in moles			
equilibrium moles			1.26
equilibrium conc value			

What is the equilibrium constant K_c for the reaction at this temperature?

8. Consider the reaction $3\ H_2(g) + N_2(g) \rightleftarrows 2\ NH_3(g)$. At 500°C, the equilibrium constant for this reaction is 6.0×10^{-2}. For each of the following situations, indicate whether or not the system is at equilibrium. If the system is not at equilibrium, indicate whether the system will shift to the right (produce more ammonia) or shift to the left (produce more hydrogen and nitrogen).

a) $(NH_3) = 2.00 \times 10^{-4}$ M $(N_2) = 1.50 \times 10^{-5}$ M $(H_2) = 0.354$ M
b) $(NH_3) = 0.0010$ M $(N_2) = 1.0 \times 10^{-5}$ M $(H_2) = 0.0020$ M
c) $(NH_3) = 1.0 \times 10^{-4}$ M $(N_2) = 5.0$ M $(H_2) = 0.010$ M

9. J. N. Spencer, G. M. Bodner, and L. H. Rickard, *Chemistry: Structure & Dynamics*, Third Edition, John Wiley & Sons, 2006. Chapter 10: Problems: 27-30, 32-42, 44, 49, 50, 54.

Problems

1. Indicate whether each of the following statements is true or false and explain your reasoning. All three statements refer to the reaction:

$$2\ SO_3(g) \leftrightarrows 2\ SO_2(g) + O_2(g)$$

   a) The rate of production of O_2 is equal to the rate of consumption of SO_3.
   b) When the above reaction reaches equilibrium, $[SO_2] = 2 \times [O_2]$.
   c) When $Q_c > K_c$, the rate of the forward reaction is greater than the rate of the reverse reaction.

2. When solutions containing $Fe^{3+}(aq)$ ions and $SCN^-(aq)$ ions are mixed together, the following equilibrium is established:

$$Fe^{3+}(aq) + SCN^-(aq) \leftrightarrows FeSCN^{2+}(aq)$$

   At equilibrium at some temperature, in 3.0 liters of total solution, there are 0.653 moles of $FeSCN^{2+}(aq)$, 0.0385 moles of $Fe^{3+}(aq)$, and 0.0465 moles of $SCN^-(aq)$.

   a) Calculate the value of the equilibrium constant, K, for the reaction at this temperature.
   b) An inquiring student pours another liter of water into the beaker holding the solution described above. She notices that the number of moles of Fe^{3+} and SCN^- are then seen to increase, and the number of moles of $FeSCN^{2+}$ decreases. Explain this observation.

3. A chemist examining the conversion of methane to other fuels was investigating the following reaction describing the reaction of methane with steam at 1200 K:

$$CH_4(g) + H_2O(g) \leftrightarrows CO(g) + 3\ H_2(g) \qquad K_c = 0.26 \text{ at } 1200\text{ K}$$

   The chemist simultaneously injected 1.8 moles of each gas (CH_4, H_2O, CO, and H_2) into a 2.00-liter flask held at 1200 K.

   a) In which direction will the reaction proceed in order to reach equilibrium? Explain your reasoning.
   b) The chemist is interested in the concentration of H_2 produced at equilibrium. Provide an algebraic expression whose solution would enable this concentration to be determined. Explicitly describe how the $[H_2]$ could be determined from the solution of this equation. DO NOT SOLVE THE EQUATION.

4. The equilibrium constants at some temperature are given for the following reactions:

$2\,NO(g) \leftrightarrows N_2(g) + O_2(g)$ $K_1 = 2.4 \times 10^{-18}$

$NO(g) + \frac{1}{2} Br_2(g) \leftrightarrows NOBr(g)$ $K_2 = 1.4$

Using this information, determine the value of the equilibrium constant, K_3, for the following reaction at the same temperature:

$\frac{1}{2} N_2(g) + \frac{1}{2} O_2(g) + \frac{1}{2} Br_2(g) \leftrightarrows NOBr(g)$

5. Sulfur dioxide reacts with molecular oxygen to form sulfur trioxide.

$2\,SO_2(g) + O_2(g) \leftrightarrows 2\,SO_3(g)$

At some temperature the equilibrium constant for this reaction is 16. If 1.0 mole of SO_2, 1.8 mole of O_2, and 4.0 mole of SO_3 are placed in a 2.0-liter flask at this temperature: a) will the reaction proceed to the left or to the right? b) set up the equation that will enable you to determine the concentrations of SO_3, O_2, and SO_3 at equilibrium (use "x"). Do not attempt to solve the equation. c) Estimate a value for x, in moles, without doing any further calculation. Briefly explain.

ChemActivity 41

The Solubility Product

(How Soluble Are Ionic Salts?)

Model 1: The Dissolution of Magnesium Hydroxide in Water.

When solid $Mg(OH)_2$ dissolves in water the chemical reaction is:

$$Mg(OH)_2(s) \leftrightarrows Mg^{2+}(aq) + 2OH^-(aq) \quad (1)$$

where $Mg^{2+}(aq)$ represents magnesium ions surrounded by water molecules.

Table 1. The results (after equilibrium has been established) of adding solid $Mg(OH)_2$ to 10.0 L of water.

Total amount of $Mg(OH)_2$ added (g)	Total amount of $Mg(OH)_2$ added (moles)	Mg^{2+} concentration in the resulting solution $[Mg^{2+}]$, (M)	OH^- concentration in the resulting solution $[OH^-]$, (M)	Mass of $Mg(OH)_2$ that does not dissolve (g)
0.00963	1.65×10^{-4}	1.65×10^{-5}	3.30×10^{-5}	0.00000
0.04815	8.26×10^{-4}	8.26×10^{-5}	1.65×10^{-4}	0.00000
0.09590	1.64×10^{-3}	1.64×10^{-4}	3.29×10^{-4}	0.00000
0.09630	1.65×10^{-3}	1.65×10^{-4}	3.30×10^{-4}	0.00000
0.09700	1.66×10^{-3}	1.65×10^{-4}	3.30×10^{-4}	0.00070
0.10000	1.71×10^{-3}	1.65×10^{-4}	3.30×10^{-4}	0.00370
0.15000	2.57×10^{-3}	1.65×10^{-4}	3.30×10^{-4}	0.05370
0.20000	3.43×10^{-3}	1.65×10^{-4}	3.30×10^{-4}	0.10370

Critical Thinking Questions

1. Choose one of the masses of $Mg(OH)_2$ in the left-hand column of Table 1 and verify that the corresponding number of moles of $Mg(OH)_2$ is correct.

2. When 8.26×10^{-4} moles of $Mg(OH)_2$ are added,

   a) why is $[Mg^{2+}] = 8.26 \times 10^{-5}$ M?

   b) why is $[OH^-] = 1.65 \times 10^{-5}$ M?

3. When $[Mg^{2+}] = 8.26 \times 10^{-5}$, $[OH^-] = 1.65 \times 10^{-4}$. Explain.

4. When 0.20000 grams of $Mg(OH)_2$ are added, how many grams dissolve?

5. When 0.09700 grams of $Mg(OH)_2$ are added, how many grams dissolve?

6. According to the table, what is the maximum number of moles of $Mg(OH)_2$ that can be dissolved in 10.0 L of water?

7. Based on your answer to CTQ 6,

   a) what is the maximum number of moles of $Mg(OH)_2$ that can be dissolved in 1.00 L of water?

   b) what is the maximum mass of $Mg(OH)_2$ that can be dissolved in 1.00 L of water?

8. Based on your answers to CTQs 6 and 7, what is the maximum value of the expression $[Mg^{2+}]\,[OH^-]^2$?

9. Is it possible for the value of $[Mg^{2+}]\,[OH^-]^2$ to be less than your answer given to CTQ 8? Explain.

Information

Once equilibrium is established between a solid material and the associated aqueous species, the solution is said to be **saturated**. For $Mg(OH)_2$, we say that the **solubility** of magnesium(II) hydroxide is 9.63×10^{-3} grams/liter or that the solubility of magnesium(II) hydroxide is 1.65×10^{-3} M.

By convention, if a saturated solution of an ionic compound is greater than about 0.1 M, we say that the compound is **soluble**. If the saturated solution is less than about 1×10^{-3} M, the compound is said to be **insoluble**. Intermediate cases are said to be moderately soluble. Experimental evidence has shown that essentially all compounds containing the nitrate ion, NO_3^-, and also all those containing the sodium ion, Na^+, are soluble in water.

Model 2: The Equilibrium Constant for $Mg(OH)_2$.

$$K_c = \frac{[Mg^{2+}(aq)]\,[OH^-(aq)]^2}{[Mg(OH)_2(s)]} \qquad (2)$$

It must be recognized that when no solid $Mg(OH)_2$ is present that $[Mg(OH)_2(s)]$ is zero and the expression for K_c is undefined. That is, the expression for K_c can only be used when solid $Mg(OH)_2$ is present.

Critical Thinking Questions

10. Show that $[Mg(OH)_2(s)] = 40.5$ mole/L. Hint: for $Mg(OH)_2(s)$, MW = 58.32 g/mole and the density of $Mg(OH)_2(s)$ is 2.36 g/mL.

11. Use Model 1 to calculate the value of $[Mg^{2+}(aq)]\,[OH^-(aq)]^2$ when solid $Mg(OH)_2$ is present.

12. Use your answers to the CTQs above to calculate the value of K_c for $Mg(OH)_2$.

13. Under what circumstances when Mg^{2+} (aq) and OH^- (aq) are present in a solution is the concentration of $Mg(OH)_2(s)$ not equal to 40.5 mole/L?

Information

Notice that for the dissolution of any solid the expression for K_c will always contain the concentration of the solid in the denominator. Because the concentration of the solid is always constant (at a particular temperature), the concentration of the solid can be combined with K_c as follows (for $Mg(OH)_2$):

$$K_c = \frac{[Mg^{2+}(aq)]\,[OH^-(aq)]^2}{[Mg(OH)_2(s)]} = 4.44 \times 10^{-13} \quad \text{(at 25°C)} \qquad (3)$$

$$K_c\,[Mg(OH)_2(s)] = [Mg^{2+}(aq)]\,[OH^-(aq)]^2 = 1.80 \times 10^{-11} = K \qquad (4)$$

This new equilibrium constant, K, is called the **solubility product**,and is given the symbol K_{sp}. The subscript "sp" is used only with an equilibrium constant, K, that describes the dissolution of a solid in water. Thus, for the dissolution of $Mg(OH)_2(s)$:

$$Mg(OH)_2(s) \leftrightharpoons Mg^{2+}(aq) + 2OH^-(aq)$$

$$K_{sp} = [Mg^{2+}(aq)]\,[OH^-(aq)]^2 = 1.80 \times 10^{-11} \quad \text{(at 25°C)} \qquad (5)$$

More Information

Furthermore, when an equilibrium expression is written for a chemical reaction, pure solids and liquids are normally omitted from the expression (here again, because they are constants), and the value of the equilibrium constant, K, is assumed to apply to the remaining species. However, whenever the equilibrium constant is denoted as K_c, all substances must be included in the equilibrium constant expression. For example:

$$PbCl_2(s) \leftrightarrows Pb^{2+}(aq) + 2Cl^-(aq) \qquad K_c = \frac{[Pb^{2+}(aq)][Cl^-(aq)]^2}{[PbCl_2(s)]} \tag{6}$$

$$PbCl_2(s) \leftrightarrows Pb^{2+}(aq) + 2Cl^-(aq) \qquad K_{sp} = [Pb^{2+}(aq)]\,[Cl^-(aq)]^2 \tag{7}$$

$$2Mg(s) + O_2(g) \leftrightarrows 2MgO(s) \qquad K_c = \frac{[MgO(s)]^2}{[Mg(s)]^2\,[O_2(g)]} \tag{8}$$

$$2Mg(s) + O_2(g) \leftrightarrows 2MgO(s) \qquad K = \frac{1}{[O_2(g)]} \tag{9}$$

Model 3: Precipitate Formation.

If a solid material forms when two solutions are mixed together, the solid is said to **precipitate** out of solution. The solid is referred to as the **precipitate**. The following represents a precipitate of $Mg(OH)_2(s)$ in equilibrium with aqueous magnesium ions and aqueous hydroxide ions.

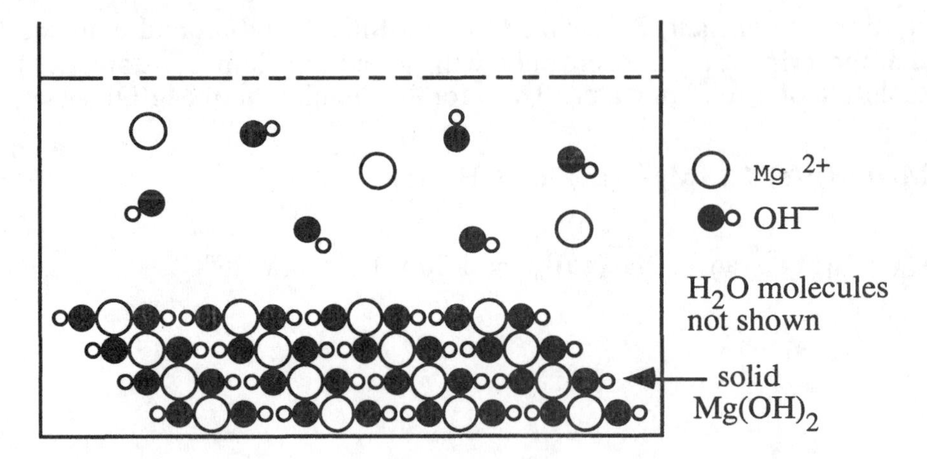

Critical Thinking Questions

14. Consider a solution formed by combining 500.0 mL of 0.12 M NaOH solution, with 500.0 mL of 0.10 M $Mg(NO_3)_2$ solution in a beaker:

a) How many moles of $Mg^{2+}(aq)$ are present (assuming that no reaction occurs)?

b) How many moles of $OH^{-}(aq)$ are present (assuming that no reaction occurs)?

c) What is the volume of the final solution?

d) At this point (assuming that no reaction occurs), what is (Mg^{2+})?

e) At this point (assuming that no reaction occurs), what is (OH^{-})?

f) Write the expression for the reaction quotient, Q_{sp}, for this reaction:

$Mg(OH)_2(s)$

Calculate the value of the reaction quotient, Q_{sp}, for the mixture described above.

g) Upon mixing, is the system at equilibrium?

If not, in what direction will it shift?

h) Does a precipitate form?

Exercises

1. Write the equilibrium constant expression, K, for each of the following reactions. Where appropriate, designate the K as K_{sp}.

   a) $2H_2(g) + O_2(g) \leftrightarrows 2H_2O(g)$

   b) $2Hg(l) + Cl_2(g) \leftrightarrows Hg_2Cl_2(s)$

   c) $BaSO_4(s) \leftrightarrows Ba^{2+}(aq) + SO_4^{2-}(aq)$

   d) $NH_4HS(s) \leftrightarrows NH_3(g) + H_2S(g)$

   e) $BaCO_3(s) \leftrightarrows BaO(s) + CO_2(g)$

   f) $NH_4Cl(s) \leftrightarrows NH_3(g) + HCl(g)$

   g) $Ag_2SO_4(s) \leftrightarrows 2Ag^+(aq) + SO_4^{2-}(aq)$

   h) $2Ag^+(aq) + SO_4^{2-}(aq) \leftrightarrows Ag_2SO_4(s)$

2. Indicate whether each of the following statements is true or false, and explain your reasoning.

   a) For the reaction $CaCO_3(s) \leftrightarrows CaO(s) + CO_2(g)$, the amount of CO_2 present at equilibrium in a 2.00 liter box is greater if 10.0 g of $CaCO_3$ are originally placed in the box than if only 5.00 g of $CaCO_3$ are originally present. (Hint: write the equilibrium constant expression, K, for the reaction.)

   b) Once equilibrium is reached, the forward and reverse chemical reactions stop.

3. A saturated solution of lithium carbonate, Li_2CO_3, is obtained after 0.0742 moles of the solid have dissolved in 1.00 liter. Calculate the value of K_{sp} for lithium carbonate.

4. What mass of MgF_2 will dissolve in 125 mL of water if $K_{sp} = 6.5 \times 10^{-9}$?

5. When 1.0 g of AgCl is placed in a beaker containing 2.00 liters of water at room temperature, only a small amount of AgCl(s) is observed to dissolve. In fact, only 8.0×10^{-5} moles of AgCl are found to dissolve.

   Calculate the equilibrium constant, K_{sp}, for the reaction:

$$AgCl(s) \leftrightarrows Ag^+(aq) + Cl^-(aq)$$

6. $PbCl_2(s)$ is not very soluble in water.

$$PbCl_2(s) \leftrightarrows Pb^{2+}(aq) + 2\ Cl^-(aq)$$

a) If x moles of $PbCl_2(s)$ dissolve in 1.00 L of water, how many moles of $Pb^{2+}(aq)$ are produced? How many moles of $Cl^-(aq)$ are produced?
b) The equilibrium constant, K_{sp}, for the dissolution of $PbCl_2(s)$ in water is 1.6×10^{-5}. What is the concentration of $Pb^{2+}(aq)$ at equilibrium? What is the concentration of $Cl^-(aq)$ at equilibrium?

7. For each of the following situations, determine whether or not a precipitate of MgF_2 is expected to form.

a) 500.0 mL of 0.050M $Mg(NO_3)_2$ is mixed with 500.0 mL of 0.010 M NaF.
b) 500.0 mL of 0.050M $Mg(NO_3)_2$ is mixed with 500.0 mL of 0.0010 M NaF.

8. Is a precipitate of $Cd(CN)_2$ expected to form when 500.0 mL of 0.010 M $Cd(NO_3)_2$ is mixed with 500.0 mL of 0.0025 M NaCN? Both cadmium(II) nitrate and sodium(I) cyanide are completely dissociated in the original solutions. The K_{sp} of $Cd(CN)_2$ is 1.0×10^{-8}.

9. J. N. Spencer, G. M. Bodner, and L. H. Rickard, *Chemistry: Structure & Dynamics*, Third Edition, John Wiley & Sons, 2006. Chapter 10: Problems: 73, 74, 76, 77, 81-84, 118, 121.

Problems

1. The K_{sp} of Ag_2SO_4 is 1.4×10^{-5}. Will a precipitate form when 250 mL of 0.12 M $AgNO_3$ is mixed with 500 mL of 0.0050 M Na_2SO_4?

2. a) Write a chemical equation that describes the dissolution of solid $AuCl_3$ to Au^{3+} and Cl^- ions found in water. b) Write the expression for the K_{sp} of $AuCl_3$. c) Calculate how many grams of Au^{3+} would be found in one liter of a saturated solution of $AuCl_3$. The K_{sp} for $AuCl_3$ is 3.2×10^{-23}.

ChemActivity 42

Acids and Bases

Model 1: Two Definitions of Acids and Bases.

Arrhenius Definitions

An **acid** is a substance that produces hydronium ions, $H_3O^+(aq)$, when it is added to water.

A **base** is a substance that produces hydroxide ions, $OH^-(aq)$, when it is added to water.

Bronsted-Lowry Definitions

An **acid** is a substance that donates a proton, H^+, to another species.

A **base** is a substance that accepts a proton, H^+, from another species.

Acid-base reactions are one of the most important types of chemical reactions.

Table 1. Equilibrium constants for some acid-base reactions.

Reaction	K_c	
$HCl(g) + H_2O(l) \rightleftarrows H_3O^+(aq) + Cl^-(aq)$	2×10^4	(1)
$NH_3(aq) + H_2O(l) \rightleftarrows NH_4^+(aq) + OH^-(aq)$	3.3×10^{-7}	(2)
$HCN(aq) + H_2O(l) \rightleftarrows H_3O^+(aq) + CN^-(aq)$	1.1×10^{-11}	(3)

Critical Thinking Questions

1. a) What chemical species are the Arrhenius acids in the forward reactions (1)-(3)?

   b) What chemical species are the Arrhenius bases in the forward reactions (1)–(3)?

   c) What chemical species are the Bronsted-Lowry acids in the forward reactions (1)–(3)?

   d) What chemical species are the Bronsted-Lowry bases in the forward reactions (1)–(3)?

2. Is it possible for a substance to act as both an acid and a base? Explain your reasoning.

3. Based on the data in Table 1, which do you think is considered the stronger acid, HCl or HCN? Explain your reasoning.

4. Consider reaction (1).
   a) What species results from the loss of a proton from the Bronsted-Lowry acid in the forward reaction?

   b) Does the species indicated in part a) (the answer that you gave) act as an acid or a base when the reverse of reaction (1) occurs?

   c) What species results from the gain of a proton by the Bronsted-Lowry base in the forward reaction?

   d) Does the species indicated in part c) act as an acid or a base when the reverse of reaction (1) occurs?

   e) Answer parts a) – d) for reactions (2) and (3) also. Describe any general relationship that you observe using a grammatically correct English sentence.

Model 2: Conjugate Pairs.

Within the Bronsted-Lowry model, certain pairs of molecules are described as a **conjugate acid-base pair**. The two species in a conjugate acid-base pair differ by a proton only. A base is said to have a conjugate acid, and an acid is said to have a conjugate base.

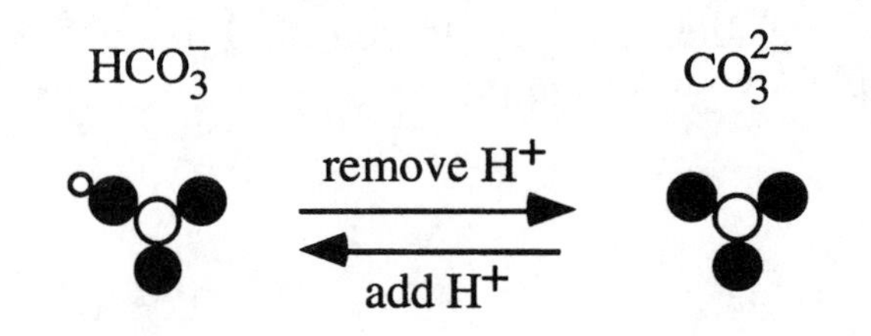

Table 2. Examples of conjugate acid-base pairs.

Acid	Base
H_2CO_3	HCO_3^-
HCO_3^-	CO_3^{2-}
H_3O^+	H_2O
H_2S	HS^-

A conjugate acid-base pair differs by a proton, H^+.
The species with more protons is the acid.

Critical Thinking Questions

5. Why is the charge on the hydrogen sulfide ion in Table 2 given as –1?

6. Answer and explain each of the following:

   a) What is the conjugate acid of NH_3?

   b) What is the conjugate base of H_2O?

   c) Define a conjugate acid-base pair.

Exercises

1. Give the conjugate base of each of the following: HSO_4^- ; HCO_3^- ; H_2O ; OH^- ; H_3O^+ ; NH_4^+ ; $CH_3NH_3^+$; HF ; CH_3COOH .

2. Give the conjugate acid of each of the following: SO_4^{2-} ; CO_3^{2-} ; H_2O ; OH^- ; O^{2-} ; NH_3 ; CH_3NH_2 ; CN^- ; CH_3COO^- ; F^- ; HCO_3^- ; NH_2^- .

3. For each of the following reactions:

$$H_2SO_4(aq) + H_2O \rightleftarrows H_3O^+(aq) + HSO_4^-(aq)$$
$$HSO_4^-(aq) + H_2O \rightleftarrows SO_4^{2-}(aq) + H_3O^+(aq)$$
$$H_2O + H_2O \rightleftarrows H_3O^+(aq) + OH^-(aq)$$
$$HCN(aq) + CO_3^{2-}(aq) \rightleftarrows HCO_3^-(aq) + CN^-(aq)$$
$$H_2S(g) + NH_3(l) \rightleftarrows HS^-(am) + NH_4^+(am)$$

(am) = dissolved in liquid ammonia

   a) Which reactant is the acid?
   b) Which reactant is the base?
   c) Find the two conjugate pairs present in the reaction.

4. Ammonia can react as an acid or a base.
   a) What is the conjugate acid of ammonia?
   b) What is the conjugate base of ammonia?
   c) Complete the following acid-base reaction in which $NH_3(l)$ acts as both an acid and a base:

$$NH_3(l) + NH_3(l) \rightarrow$$

5. Complete the following table of conjugate acids and bases:

Acid	Base
H_2S	
	S^{2-}
	NO_2^-
H_3PO_4	
	OCN^-
H_3O^+	
OH^-	
	F^-
	HPO_4^{2-}
HOCl	

6. J. N. Spencer, G. M. Bodner, and L. H. Rickard, *Chemistry: Structure & Dynamics*, Third Edition, John Wiley & Sons, 2006. Chapter 11: Problems: 5-7, 9, 15, 16, 19-22, 156.

ChemActivity 43

Strong and Weak Acids

(How Strong Is an Acid?)

Model 1: Strong and Weak Acids.

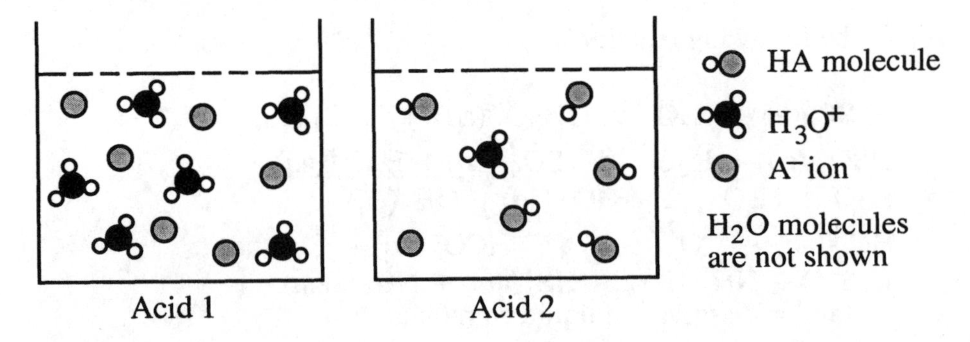

A **strong acid** is one that is essentially 100% dissociated in water; that is, when 1.0 mole of the acid is added to enough water to make 1.0 liter of solution, the resulting $[H_3O^+]$ is essentially 1.0 M.

A **weak** acid is one that is significantly less than 100% dissociated in water; that is, when 1.0 mole of a weak acid is added to enough water to make 1.0 liter of solution, the resulting $[H_3O^+]$ is significantly less than 1.0 M.

Critical Thinking Questions

1. Is Acid 1, shown in Model 1, a strong acid or a weak acid?

2. Is Acid 2, shown in Model 1, a strong acid or a weak acid?

Model 2: K_c and K_a for Acids.

When an acid HA is placed in water, hydronium ions are produced according to the reaction

$$HA(aq) + H_2O \rightleftarrows H_3O^+(aq) + A^-(aq)$$

The equilibrium constant, K_c, for this type of reaction is:

$$K_c = \frac{[H_3O^+]\ [A^-]}{[HA]\ [H_2O]} \qquad (1)$$

Most solutions are sufficiently dilute that the concentration of water is the same before and after reaction with the acid. The concentration of the water is incorporated into the value of K_c and the equilibrium expression is given a special name and symbol—the acid-dissociation constant, K_a.

$$[H_2O] \approx \frac{1000\ g/L}{18\ g/mol} \approx 55\ M \tag{2}$$

$$K_c = \frac{[H_3O^+]\ [A^-]}{[HA]\ [55]} \tag{3}$$

$$K_c \times 55 = K_a = \frac{[H_3O^+]\ [A^-]}{[HA]} \tag{4}$$

Table 1. The names and formulas of acids commonly encountered in general chemistry courses and the values of K_c and K_a for each.

Acid Name	Molecular Formula	K_c	K_a
hydroiodic acid	HI	5×10^7	3×10^9
hydrobromic acid	HBr	2×10^7	1×10^9
hydrochloric acid	HCl	2×10^4	1×10^6
perchloric acid	$HOClO_3$	2×10^6	1×10^8
sulfuric acid	H_2SO_4	2×10^1	1×10^3
hydronium ion	H_3O^+	1	55
nitric acid	HNO_3	≈ 0.5	28
acetic acid	CH_3COOH	3.2×10^{-7}	1.75×10^{-5}
carbonic acid	H_2CO_3	8.2×10^{-9}	4.5×10^{-7}
hydrofluoric acid	HF	1.3×10^{-5}	7.2×10^{-4}
hydrosulfuric acid	H_2S	1.8×10^{-9}	1.0×10^{-7}
nitrous acid	HNO_2	9.2×10^{-6}	5.1×10^{-4}
phosphoric acid	H_3PO_4	1.3×10^{-4}	7.1×10^{-3}
water	H_2O	3.3×10^{-18}	1.8×10^{-16}

The relative strength of an acid is determined by the relative H_3O^+ concentration produced at equilibrium for a given molarity of the acid. For example, if a 0.5 M solution of HA has $[H_3O^+] = 1 \times 10^{-4}$ M and a 0.5 M solution of HX has $[H_3O^+] = 1 \times 10^{-3}$ M, then HX is a **stronger** acid than HA (even though both are considered to be **weak** acids).

Critical Thinking Questions

3. The K_a for acetic acid, given in Table 1, is 1.75×10^{-5}. Show that the value given for K_c of acetic acid, given in Table 1, is correct.

4. In a solution of nitrous acid: $[HNO_2] = 1.33$ M; $[H_3O^+] = 0.026$ M; $[NO_2^-] = 0.026$ M. Show that K_a is correct in Table 1.

k

5. Nitric acid and the acids above it in Table 1 are considered to be strong acids. Explain why.

6. Acetic acid and the acids below it in Table 1 are considered to be weak acids. Explain why.

7. Of the weak acids in Table 1 (ignore H_2O), which one will produce the highest $[H_3O^+]$ in a solution of a given molarity of acid? Explain your reasoning. (You should answer this question without doing extensive equilibrium calculations.)

8. Of the weak acids in Table 1 (ignore H_2O), which one will produce the lowest $[H_3O^+]$ in a solution of a given molarity of acid? Explain your reasoning. (You should answer this question without doing extensive equilibrium calculations.)

9. Are all of the **strong** acids of equal strength? If not, what is the strongest strong acid? The weakest strong acid?

Exercises

1. What are the names and the chemical formulas for the conjugate bases of the acids listed in Table 1?

2. Rank the weak acids in Table 1 in order from strongest to weakest.

3. Write the balanced chemical equation for the reaction of HF with water. What is the expression for K_a? Choose three additional acids from Table 1 and provide the chemical reaction with water and the expression for K_a.

4. Indicate whether the following statement is true or false and explain your reasoning:

   A 0.25 M solution of acetic acid has a higher $[H_3O^+]$ than does a 0.25 M solution of nitrous acid.

5. J. N. Spencer, G. M. Bodner, and L. H. Rickard, *Chemistry: Structure & Dynamics*, Third Edition, John Wiley & Sons, 2006. Chapter 11: Problems: 48, 49, 51, 53, 139ab, 155.

Model 3: Neutral, Acidic, and Basic Solutions.

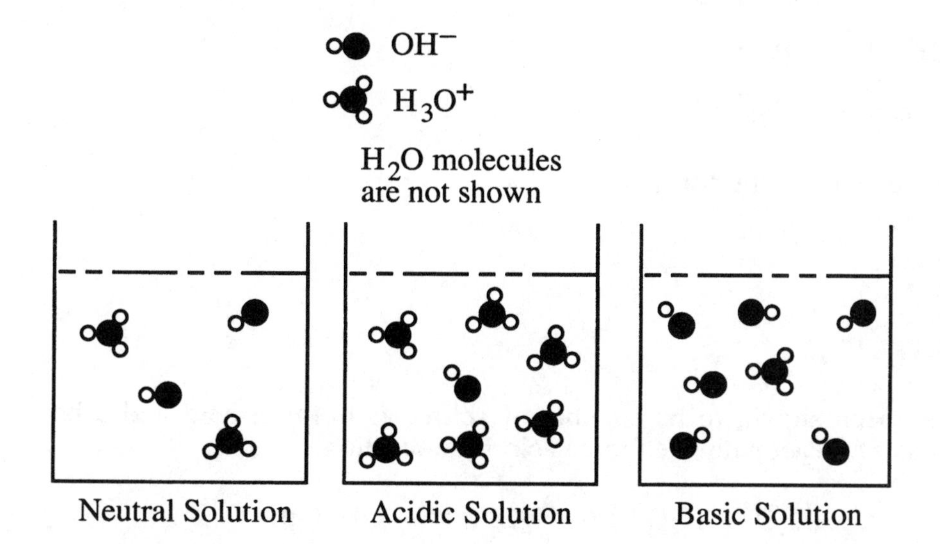

Aqueous solutions in which there is an excess of [H_3O^+] (with respect to [OH^-]) are said to be **acidic**. Solutions in which there is an excess of [OH^-] (with respect to [H_3O^+]) are said to be **basic**. Solutions in which [H_3O^+] = [OH^-] are said to be **neutral**.

Table 2. Characteristics of solutions of various amounts of hypochlorous acid, HOCl, dissolved in water to make 1.00 L of solution at 25 °C.

Moles of HOCl added	[H_3O^+] (M)	[OH^-] (M)
0.00	1.0×10^{-7}	1.0×10^{-7}
0.30	9.3×10^{-5}	1.1×10^{-10}
0.75	1.5×10^{-4}	6.8×10^{-11}
1.00	1.7×10^{-4}	5.9×10^{-11}

Critical Thinking Questions

10. Which solution(s) in Table 2 are acidic? Which are neutral? Which are basic?

11. According to the data in Table 2, is HOCl an acid or a base? Explain your answer.

12. Use Table 2. Which, if any, of these expressions is a constant?

    a) $[H_3O^+] + [OH^-]$

    b) $\dfrac{[H_3O^+]}{[OH^-]}$

    c) $[H_3O^+] - [OH^-]$

    d) $[H_3O^+] \times [OH^-]$

    What is the value of the constant?

Information

Water has been shown to be capable of acting as both an acid and a base; it is possible for water to react with itself in an acid-base reaction.

$$H_2O + H_2O \rightleftarrows H_3O^+(aq) + OH^-(aq)$$

13. Is "pure" water considered to be acidic, basic, or neutral? Explain your reasoning.

14. a) Write the expressions for the K_c and the K_a of water; see equations (1-4).

    b) Table 1 gives the value of the K_a of water, 1.8×10^{-16}. Recall that $[H_2O] \approx$ 55 M. Use this information and the expression for the K_a of water to calculate $[H_3O^+] \times [OH^-]$.

    c) How well does this value for $[H_3O^+] \times [OH^-]$ agree with the value determined in CTQ 11?

15. The value of $[H_3O^+] \times [OH^-]$ is given a special name and symbol—the water-dissociation equilibrium constant, K_w. The accepted value of K_w at 25 °C is 1.0×10^{-14}. How well does this value compare to your answers above?

16. Recall that at 25°C, $K_w \equiv [H_3O^+] \times [OH^-] = 1.0 \times 10^{-14}$. Calculate the hydronium ion concentration and the hydroxide ion concentration in "pure" water.

Exercises

6. The hydronium ion concentration of a sample of lemon juice at 25°C is 6.3×10^{-3} M. What is the hydroxide ion concentration?

7. The hydroxide ion concentration of a sample of vinegar at 25°C is 3.3×10^{-12} M. What is the hydronium ion concentration?

8. For each of the following hydronium ion concentrations, what is the hydroxide ion concentration? 3.5×10^{-5} M ; 7.1×10^{-1} M ; 4.5×10^{-10} M ; 2.1×10^{-7} M.

9. For each of the following hydroxide ion concentrations, what is the hydronium ion concentration? 3.5×10^{-5} M ; 7.1×10^{-10} M ; 5.7×10^{-12} M; 1.1×10^{-7} M.

10. Which of the solutions in Ex. 8 are acidic? Which are basic?

11. Which of the solutions in Ex. 9 are acidic? Which are basic?

12. Indicate whether the following statements are true or false and explain your reasoning:
    a) All acidic solutions have $[OH^-] < 10^{-7}$ M.
    b) A solution is considered to be acidic whenever $[H_3O^+] > 0$.

13. J. N. Spencer, G. M. Bodner, and L. H. Rickard, *Chemistry: Structure & Dynamics*, Third Edition, John Wiley & Sons, 2006. Chapter 11: Problems: 26, 27, 141.

ChemActivity 44

Weak Acid/Base Dissociation

(How Much Acid or Base Reacts?)

Model 1: A Weak Acid Increases the Hydronium Concentration of a Solution, but the Amount of Dissociation is Small.

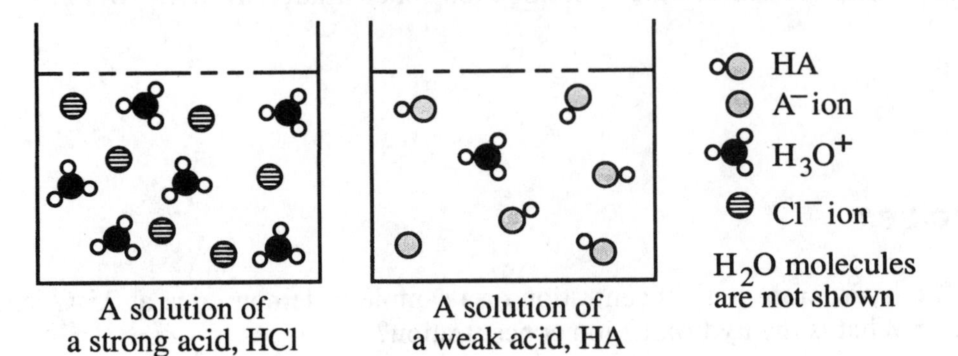

A solution of a strong acid, HCl

A solution of a weak acid, HA

Table 1. Characteristics of solutions of various amounts of hypochlorous acid, HOCl, dissolved in water to make 1.00 L of solution at 25°C.

Moles of HOCl added	$[H_3O^+]$ (M)	$[OH^-]$ (M)
0.00	1.0×10^{-7}	1.0×10^{-7}
0.30	9.3×10^{-5}	1.1×10^{-10}
0.75	1.5×10^{-4}	6.8×10^{-11}
1.00	1.7×10^{-4}	5.9×10^{-11}

Critical Thinking Questions

1. Complete and balance the following reaction:

$$HOCl(aq) + H_2O(l) \rightleftarrows H_3O^+ + OCl^-$$

Write the equilibrium expression for the K_a of HOCl.

$$\frac{[H_3O^+][Cl^-]}{[HOCl]}$$

2. Complete the following table assuming that 0.30 moles of HOCl are added to sufficient water to make 1.0 L of solution at 25°C:

	HOCl	H_3O^+	OCl^-
initial moles	0.30	0	0
change in moles	$-x$	$+x$	$+x$
equilibrium moles	$0.30 - x$	x	x
equilibrium conc	$\frac{0.30 - x}{1.0}$	x	x

3. The K_a of HOCl, at 25°C, is 2.9×10^{-8}. Find "x" (CTQ 2), and enter the equilibrium concentration values in the last row of the following table.

	HOCl	H_3O^+	OCl^-
initial moles	0.30	0	0
change in moles			
equilibrium moles			
equilibrium conc			
equilibrium conc value			

Verify that your equilibrium concentrations are correct.

4. Add the missing values for [HOCl] and [OCl^-] to Table 1a.

Table 1a. Characteristics of solutions of various amounts of hypochlorous acid, HOCl, dissolved in water to make 1.00 L of solution at 25°C.

Moles of HOCl added	[H_3O^+] (M)	[OH^-] (M)	[HOCl] (M)	[OCl^-] (M)
0.00	1.0×10^{-7}	1.0×10^{-7}	0	0
0.30	9.3×10^{-5}	1.1×10^{-10}		
0.75	1.5×10^{-4}	6.8×10^{-11}	0.75	1.5×10^{-4}
1.00	1.7×10^{-4}	5.9×10^{-11}	1.0	1.7×10^{-4}

5. Explain why the number of moles of HOCl added is equal to the number of moles of HOCl at equilibrium in spite of the fact that some of the HOCl reacts!

6. Explain why the equilibrium concentration of H_3O^+ is equal to the equilibrium concentration of OCl^-.

Model 2: A Weak Base Increases the Hydroxide Concentration of a Solution, but the Amount of Reaction Is Small.

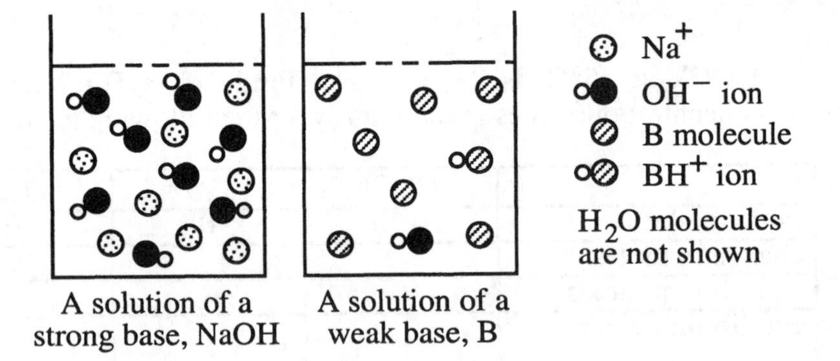

When a base B is placed in water, hydroxide ions are produced according to the reaction

$$B(aq) + H_2O \rightleftarrows BH^+(aq) + OH^-(aq)$$

The equilibrium constant, K_c, for this type of reaction is:

$$K_c = \frac{[BH^+]\,[OH^-]}{[B]\,[H_2O]}$$

Most solutions are sufficiently dilute that the concentration of water is the same before and after reaction with the base. The concentration of the water is incorporated into the value of K_c and the equilibrium expression is given a special name and symbol—the base-dissociation constant, K_b.

$$K_c = \frac{[BH^+]\,[OH^-]}{[B]\,[55]}$$

$$K_c \times 55 = K_b = \frac{[BH^+]\,[OH^-]}{[B]}$$

Table 2. Characteristics of solutions of various amounts of the base pyridine, $C_5H_5N(l)$ dissolved in water to make 1.00 L of solution at 25 °C.

Moles of C_5H_5N added	$[H_3O^+]$ (M)	$[OH^-]$ (M)
0.00	1.0×10^{-7}	1.0×10^{-7}
0.30	4.4×10^{-10}	2.3×10^{-5}
0.75	2.8×10^{-10}	3.6×10^{-5}
1.00	2.4×10^{-10}	4.1×10^{-5}

Critical Thinking Questions

7. How was the species BH^+, o⊘, produced in Model 2?

8. Determine the value of K_w from the data given in Table 2. Does this value agree with your value calculated from Table 2 of **ChemActivity 43: Strong and Weak Acids**?

9. Complete and balance the following reaction for the weak base pyridine:

$$C_5H_5N(aq) + H_2O \rightleftarrows$$

Write the equilibrium expression for the K_b of C_5H_5N.

10. Complete the following table assuming that 0.30 moles of C_5H_5N are added to sufficient water to make 1.0 L of solution at 25°C:

	C_5H_5N	OH^-	$C_5H_5NH^+$
initial moles	0.30	0	0
change in moles	$-x$		
equilibrium moles			
equilibrium conc			

11. The K_b of C_5H_5N, at 25°C, is 1.7×10^{-9}. Find "x" (CTQ 10), and enter the equilibrium concentration values in the last row of the following table.

	C_5H_5N	OH^-	$C_5H_5NH^+$
initial moles	0.30	0	0
change in moles			
equilibrium moles			
equilibrium conc			
equilibrium conc value			

Verify that your equilibrium concentrations are correct.

12. Add the missing values for $[C_5H_5N]$ and $[C_5H_5NH^+]$ to Table 2a.

Table 2a. Characteristics of solutions of various amounts of the base pyridine, $C_5H_5N(l)$, dissolved in water to make 1.00 L of solution at 25°C.

Moles of C_5H_5N added	$[H_3O^+]$ (M)	$[OH^-]$ (M)	$[C_5H_5N]$ (M)	$[C_5H_5NH^+]$ (M)
0.00	1.0×10^{-7}	1.0×10^{-7}	0	0
0.30	4.4×10^{-10}	2.3×10^{-5}		
0.75	2.8×10^{-10}	3.6×10^{-5}	0.75	3.6×10^{-5}
1.00	2.4×10^{-10}	4.1×10^{-5}	1.00	4.1×10^{-5}

13. Explain why the moles of C_5H_5N added are equal to the moles of C_5H_5N at equilibrium in spite of the fact that some of the C_5H_5N reacts!

14. Explain why the equilibrium concentration of OH^- is equal to the equilibrium concentration of $C_5H_5NH^+$.

Model 3: A Simple Equation to Describe a Solution of a Weak Acid.

When an acid HA is placed in water, hydronium ions are produced according to the reaction

$$HA(aq) + H_2O \rightleftarrows H_3O^+(aq) + A^-(aq)$$

The acid-ionization constant expression is

$$K_a = \frac{[H_3O^+]\ [A^-]}{[HA]}$$

Typically[1], for weak acids

$$[H_3O^+] \approx [A^-] \text{ and} \tag{1}$$

$$[HA] = (HA)_o - [A^-] \approx [HA]_o \tag{2}$$

$$K_a = \frac{[H_3O^+]\ [A^-]}{[HA]} \approx \frac{[H_3O^+]^2}{(HA)_o - [A^-]} \tag{3}$$

$$K_a \approx \frac{[H_3O^+]^2}{(HA)_o} \tag{4}$$

Read in your textbook, or other materials assigned by your instructor, for more information about when these approximations are valid. One source is J. N. Spencer, G. M. Bodner, and L. H. Rickard, *Chemistry: Structure & Dynamics*, Third Edition, John Wiley & Sons, 2006, Section 10.8.

Critical Thinking Questions

15. For a weak acid, why is $[H_3O^+] = [A^-]$?

16. For a weak acid, why is $[HA] = (HA)_o$?

[1]Equation (1) will not be true for a solution of an extremely dilute acid. This situation will rarely be encountered in this course.

17. A 0.50 M solution of CH_3COOH has $[CH_3COO^-] = 3.0 \times 10^{-3}$ M. What are the values of:

$(CH_3COOH)_o$ =
$[CH_3COOH]$ =
$[H_3O^+]$ =
K_a =

18. A 0.50 M solution of hydrozoic acid contains 3.1×10^{-3} M H_3O^+. Use equation (3) to calculate the value of K_a for hydrozoic acid. Use equation (4) to calculate the value of K_a for hydrozoic acid. Does it matter if equation (3) or equation (4) is used? Explain.

19. Derive expressions analogous to equations (1)–(4) for a weak base.

Exercises

1. The hydronium ion concentration of a 0.30 M solution of a weak acid is 5.7×10^{-4} M. What is the value of K_a for this acid?

2. The hydroxide ion concentration of a 0.200 M solution of a weak acid is 7.0×10^{-10} M. What is the value of K_a for this acid?

3. What is the equilibrium concentration of the weak acid HONO in a 0.80 M solution of HONO? Write the chemical reaction of HONO with water. Write the expression for K_a. What is the numerical value of the hydronium ion concentration?

4. What is the equilibrium concentration of the weak base CH_3NH_2 in a 1.5 M solution of CH_3NH_2 ? Write the chemical reaction of CH_3NH_2 with water. Write the expression for K_b. What is the numerical value of $[OH^-]$?

5. J. N. Spencer, G. M. Bodner, and L. H. Rickard, *Chemistry: Structure & Dynamics*, Third Edition, John Wiley & Sons, 2006. Chapter 11: Problems: 80, 82, 85-87, 100-102, 140.

Problems

1. The sting from some ant bites is due to formic acid, HCOOH. When 0.10 moles of formic acid are dissolved in enough water to make 1.00 liter of solution, the resulting $[H_3O^+] = 0.0042$ M. The reaction that occurs is

$$HCOOH(aq) + H_2O(l) \leftrightarrows HCOO^-(aq) + H_3O^+(aq)$$

a) What is $[OH^-]$ in this solution?
b) Draw the Lewis structure of formic acid.
c) Draw the Lewis structure of the conjugate base of formic acid.
d) What is the value for K_a for formic acid? Show all of your work.

ChemActivity 45

pH

(What Is pH Good pHor?)

Model: pH Is Defined as – log[H_3O^+].

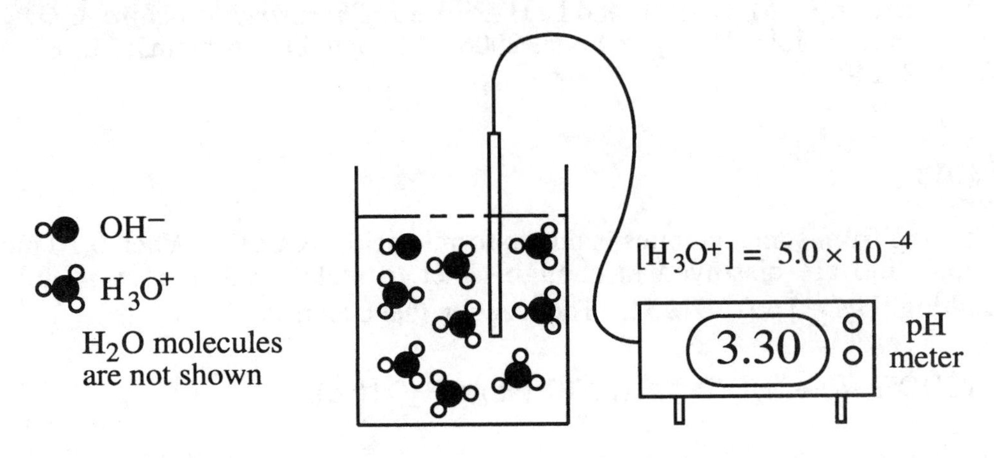

The water-dissociation equilibrium constant, K_w, is (at 25°C)

$$K_w = [H_3O^+]\,[OH^-] = 1.0 \times 10^{-14}$$

Several definitions have been found to be useful:

$$pH \equiv -\log\,[H_3O^+]$$

$$pOH \equiv -\log\,[OH^-]$$

$$pK_w \equiv -\log K_w$$

In general, $pX \equiv -\log X$

For pX expressions involving concentrations, the concentration units are always mole/liter, but they are omitted in the calculation. Thus, for example, values for pH are unitless.

Note on significant figures. For logarithms, the number of significant figures is determined by the number of digits to the right of the decimal point. The value to the left of the decimal point indicates only the power of ten by which the number is to be multiplied. For example, if pH = 2.15, then the corresponding value for [H_3O^+] has only 2 significant figures (7.1×10^{-3}). The "2" in the value "2.15" is not considered a significant figure in the conversion to concentration.

Critical Thinking Questions

1. Show that the pH reading in the model is correct (given that the hydronium ion concentration is 5.0×10^{-4} M).

2. Consider a neutral aqueous solution:

   a) What is the pH of a neutral aqueous solution?

   b) What is the pOH of a neutral aqueous solution?

3. What values of pH characterize:

   a) an acidic solution?

   b) a basic solution?

4. What is the numerical value of pK_w?

5. Recall that log (A × B) = log A + log B. What is the relationship between pH, pOH, and pK_w?

6. For a weak acid, HA:

$$HA(aq) + H_2O \rightleftarrows H_3O^+(aq) + A^-(aq)$$

   a) Write the equilibrium expression for K_a.

   b) Show that $pH = pK_a - \log \frac{[HA]}{[A^-]}$

$\log(A \cdot B) = -\log A + -\log B$

$pH = pKa - \log[HA]/[A^-]$

$\frac{[HA]}{[A^-]} \cdot Ka = \frac{[H_3O^+][A^-]}{[HA]} \cdot \frac{[HA]}{[A^-]}$

$Ka \frac{[HA]}{[A^-]} = [H_3O^+]$

$-\log Ka \frac{[HA]}{[A^-]} = -\log[H_3O^+]$

$-\log Ka + -\log \frac{[HA]}{[A^-]} = pH$

$pKa - \log\left(\frac{[HA]}{[A^-]}\right) = pH$

Exercises

1. Determine the pH and pOH of each of the following solutions, and indicate whether each is acidic, basic, or neutral:

   a) Milk, $[H_3O^+] = 3.2 \times 10^{-7}$ M
   b) Pickle juice , $[H_3O^+] = 2.0 \times 10^{-4}$ M
   c) Beer, $[H_3O^+] = 3.2 \times 10^{-5}$ M
   d) Blood, $[H_3O^+] = 4.0 \times 10^{-8}$ M

2. Determine the $[H_3O^+]$ and $[OH^-]$ of each of the following solutions, and indicate whether each is acidic, basic, or neutral:

   a) Lime juice, pH = 1.9
   b) Tomato juice, pH = 4.2
   c) Saliva, pH = 7.0
   d) Kitchen cleanser, pH = 9.3

3. Rank the following aqueous solutions in order of increasing pH without referring to a table of acid/base constants. Explain your reasoning.

   a) pure H_2O
   b) x molar NaOH
   c) x molar HCl
   d) x molar acetic acid

4. Rank the following aqueous solutions in order of increasing pH (you will need to refer to a table of acid/base constants). Explain your reasoning.

   a) pure H_2O
   b) x molar NaOH
   c) x molar C_5H_5N
   d) x molar NH_3

5. The pH of a 0.040 M solution of HOBr is 5.01. Determine K_a for the weak acid HOBr.

6. The pH of 0.300 M formic acid is 2.13. What is the K_a of formic acid?

7. The pH of a 0.15 M hydrazine (H_2NNH_2; a weak base) is 10.68. What is the K_b of hydrazine ?

For the following problems, you will need to refer to a table of acid/base constants.

8. How many moles of NH_3 must be dissolved in 1.00 liters of aqueous solution to produce a solution with a pH of 11.47?

9. What are the pH and the pOH of 0.125 M $HClO_4$?

10. What are the pH and the pOH of 0.125 M HCl?

11. What are the H_3O^+ and OH^- concentrations in a 125 mL solution prepared with 0.100 mol of HI and water?

12. What are the H_3O^+ and OH^- concentrations in a 125 mL solution prepared with 0.100 mol of NaOH and water?

13. What is the pH of a solution prepared from 6.50 g of benzoic acid (a weak acid), C_6H_5COOH, and 500 mL of water?

14. What is the pH of a solution prepared from 3.52 g of aniline (a weak base), $C_6H_5NH_2$, and 200 mL of water?

15. Calculate the pH for each of the following solutions:

    a) 0.45 M NaOH
    b) 0.45 M HCl
    c) 0.45 M CH_3COOH
    d) 0.45 M CH_3NH_2

16. J. N. Spencer, G. M. Bodner, and L. H. Rickard, *Chemistry: Structure & Dynamics*, Third Edition, John Wiley & Sons, 2006. Chapter 11: Problems: 34-36, 40, 74, 75, 79, 103, 132, 136.

Problem

1. When 500.0 mL of 0.10 M NaOH solution (containing Na^+ and OH^- ions) is mixed with 500.0 mL of 0.10 M $Mg(NO_3)_2$ solution (containing Mg^{2+} and NO_3^- ions) a precipitate of solid $Mg(OH)_2$ forms, and the resulting aqueous solution has pH = 9.43. Based on this information, determine the value of K_{sp} for $Mg(OH)_2$. Show your reasoning clearly.

ChemActivity 46

Relative Acid Strength

(What Makes an Acid Strong?)

Model 1: Bond Strengths of Two Acids With Related Structure.

H—Ö—H (with two lone pairs on O)   H—S̈—H (with two lone pairs on S)

O–H bond enthalpy = 463 kJ/mol   S–H bond enthalpy = 367 kJ/mol

Critical Thinking Questions

1. Based on the information in Model 1, which bond is easier to break: O–H or S–H?

2. Explain the relative bond strengths from CTQ 1 in terms of the molecular structure of the two molecules in Model 1.

3. The H–X bond strength decreases in the series HF, HCl, HBr, HI. Explain this in terms of the molecular structure of the molecules.

4. Predict which is most likely to be the stronger acid: H_2O or H_2S. Explain your reasoning.

5. Based on your analysis above, which do you expect to be the stronger acid, NH_4^+ or PH_4^+? Explain your reasoning.

Model 2: Relative Acid Strength for Molecules with Similar Structures but Very Different H-Q Bond Strengths.

Table 1. Characteristics of some acids. Acids within each group have similar structures.

Similar Structures	Acid	Bond	Bond Enthalpy (kJ/mole)	K_a
A	H_2O	H–O	463	1.8×10^{-16}
A	H_2S	H–S	367	1.0×10^{-7}
B	NH_4^+	H–N	390	5.6×10^{-10}
B	PH_4^+	H–P	325	$\approx 10^{14}$
C	HF	H–F	568	7.2×10^{-4}
C	HCl	H–Cl	432	1×10^{6}
C	HBr	H–Br	366	1×10^{9}
C	HI	H–I	298	3×10^{9}

Critical Thinking Questions

6. Based on the results presented in Table 1, which of the following statements best describes relative acid strength? Explain your reasoning.

- When the bond strengths between the acidic hydrogen and the atom to which it is attached are not comparable, the acid strength increases as the bond strength increases.
- When the bond strengths between the acidic hydrogen and the atom to which it is attached are not comparable, the acid strength decreases as the bond strength increases.

Model 3: Acidity of Molecules with More Than One Hydrogen Atom.

For molecules with more than one hydrogen atom, the hydrogen atom with the largest partial positive charge tends to be the acidic hydrogen.

Figure 1. Partial charges on the atoms in acetic acid and trichloroacetic acid.

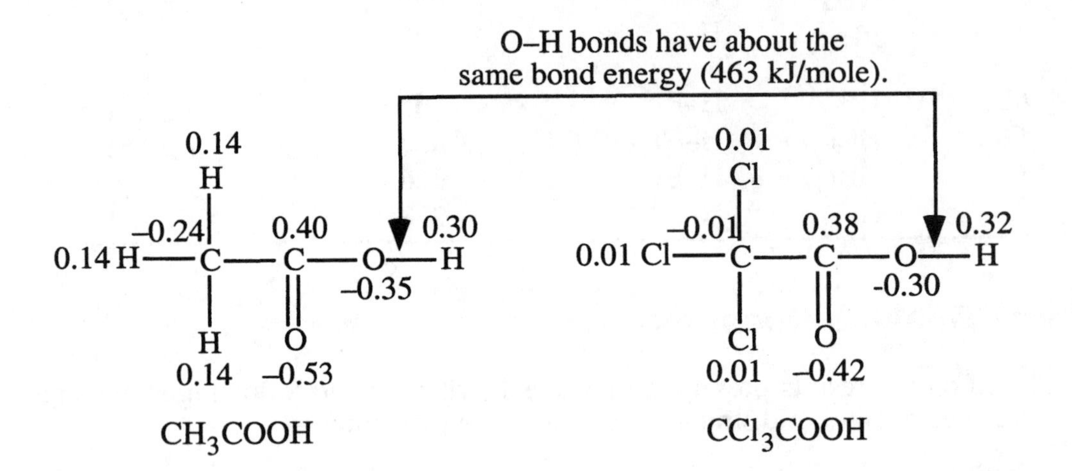

Critical Thinking Questions

7. What is the sum of the partial charges on each of the molecules in Figure 1?

8. a) Which hydrogen in CH_3COOH is the acidic hydrogen?

   b) Which hydrogen in CCl_3COOH is the acidic hydrogen?

9. Predict which is most likely to be the stronger acid: CH_3COOH or CCl_3COOH. Explain your reasoning.

Model 4: The Acidity of Molecules that Contain O-H Bonds.

> For molecules that contain one or more oxygen atoms, the most acidic hydrogen atom is always bonded to an oxygen atom.

10. Which hydrogen atom is the most acidic hydrogen atom in CH_3COOH?

11. Which hydrogen atom is the most acidic hydrogen atom in CH_3CH_2OH?

12. Explain why Model 4 is consistent with Model 3.

Model 5: Relative Acid Strength for Molecules with Similar Structures, X–Q–H, and the Q Atom is Held Constant.

Table 2. Characteristics of some acids. Acids within each group have similar structures.

Similar Structures	Q atom	Acid	Partial Charge on Acidic Hydrogen (MOPAC; Water Solvent)	K_a
A	O	CH_3COOH	0.298	1.8×10^{-5}
A	O	$CH_2ClCOOH$	0.308	1.4×10^{-3}
A	O	$CHCl_2COOH$	0.317	5.1×10^{-2}
A	O	CCl_3COOH	0.325	0.22
B	S	HOCl	0.280	2.9×10^{-8}
		HOBr	0.275	2.4×10^{-9}
B	S	HOI	0.270	2.3×10^{-11}

Critical Thinking Questions

13. Why does the partial charge on the acidic hydrogen increase in the series CH_3COOH, $CH_2ClCOOH$, $CHCl_2COOH$, CCl_3COOH?

14. What structural feature accounts for the decrease in the partial charge on the acidic hydrogen in the series HOCl, HOBr, HOI?

15. Are the results presented in Table 2 consistent with one or both of the following statements? Explain your reasoning.

- When the bond strengths between the acidic hydrogen and the atom to which it is attached are roughly comparable, the acid strength increases as the partial positive charge on the acidic hydrogen increases.
- When the bond strengths between the acidic hydrogen and the atom to which it is attached are roughly comparable, the acid strength increases as the partial positive charge on the acidic hydrogen decreases.

16. Predict the value of K_a for CF_3COOH. Explain your reasoning.

17. a) In what way (or ways) are the acids in Table 1 all structurally similar?

b) In what way (or ways) are the acids in Table 2 all structurally similar?

c) How are the acids in Table 1 structurally different than the acids in Table 2?

18. a) For each of the three groups of acids in Table 1, use electronegativities to predict the ordering of the partial positive charge on the acidic H atom. Explain your reasoning.

b) If partial positive charge on hydrogen was the most important factor in determining relative acid strength, which acid in each of the three groups of acids in Table 1 would be the strongest? Is this consistent with the relative acid strengths given in Table 1?

c) Based on the data in Table 1 (which compares acidic hydrogen atoms bonded to different atoms) which factor is most important in determining relative acid strength within each group in Table 1: bond strength or partial charge on H? Explain your reasoning.

Exercises

1. For each of the following pairs of acids, predict which will have the larger value of K_a, and explain your reasoning.

a)	H_2S	and	H_2Se
b)	HONO	and	HOPO
c)	NH_4^+	and	Cl_3NH^+
d)	$(HO)_2SeO_2$	and	$(HO)_2SO_2$
e)	H_2S	and	H_2Te
f)	$HONO_2$	and	HONO

2. Rank the following solutions in order of increasing pH. Explain your reasoning.

   x M HBr
   x M CH_3COOH
   x M CF_3COOH
   x M KBr
   x M NH_3

3. Consider the relative acid strengths of H_2O and HF. Which of the two factors dominates the determination of relative acidity for these acids? Why is this the case?

4. J. N. Spencer, G. M. Bodner, and L. H. Rickard, *Chemistry: Structure & Dynamics*, Third Edition, John Wiley & Sons, 2006. Chapter 11: Problems: 66, 67, 69.

Problems

1. Aspirin (shown below) is a weak acid with $K_a = 3.0 \times 10^{-4}$. a) Complete the following reaction of aspirin with water. b) Calculate the pH of 50 mL of 0.15 M aspirin.

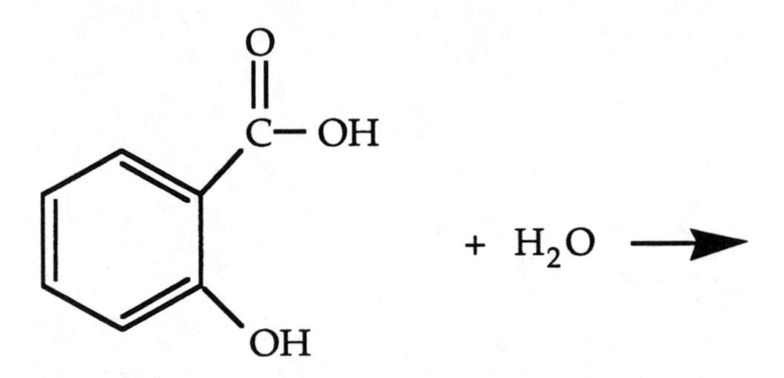

2. Predict which will have the larger value for K_a and give an explanation: $HOIO_3$ or $HSIO_3$?

ChemActivity 47

Acid/Base Strength of Conjugate Pairs

(How Are K_a and K_b Related?)

Model 1: The Mathematical Relationship between K_a and K_b of a Conjugate Pair.

Acid	K_a	Conjugate Base	K_b	$K_a \times K_b$
HF hydrofluoric acid	$\frac{[H_3O^+][F^-]}{[HF]}$	F^- fluoride ion	$\frac{[OH^-][HF]}{[F^-]}$	
HONO +H2O nitrous acid	[H3O] [ONO] / [HONO	ONO	[OH) [HONO] / [ONO]	
NH4+H2O	[H3O) [N-]	NH_3 ammonia		

Critical Thinking Questions

1. Fill in the missing entries in Model 1.

2. For each acid and each conjugate base in Model 1, write the balanced chemical reaction that has the K_a or K_b as its equilibrium constant. The conjugate base F^- is worked as an example:

$$F^-(aq) + H_2O(l) \rightleftarrows HF(aq) + OH^-(aq) \qquad K_b = \frac{[OH^-][HF]}{[F^-]}$$

3. For each of the following, describe the common features:

   a) all K_a expressions

   b) all K_b expressions

   c) all $K_a \times K_b$ products

4. Provide an expression relating K_w to K_a and K_b of a conjugate acid-base pair.

5. Describe how to determine the value of K_b for a base—given the value of K_a for its conjugate acid.

6. Consider two acids, HA and HX, with HA being a stronger acid than HX.

   a) Which acid has a larger value of K_a?

   b) Which conjugate base, A^- or X^-, has a larger value of K_b?

   c) Provide a qualitative description of the relationship between the relative strength of an acid and the relative strength of its conjugate base.

Exercises

1. Give the conjugate base and the K_b of the conjugate base for each of the acids in the table below:

Acid	K_a	Base	K_b
CH_3COOH	1.8×10^{-5}		
H_2CO_3	4.5×10^{-7}		
H_2S	1.0×10^{-7}		
HNO_2	5.1×10^{-4}		
NH_4^+	5.6×10^{-10}		

2. Give the conjugate acid and the K_a of the conjugate acid for each of the bases in the table below:

Base	K_b	Acid	K_a
NH_3	1.8×10^{-5}		
CH_3COO^-	5.6×10^{-10}		
$C_6H_5NH_2$	4.0×10^{-10}		
NO_2^-	2.0×10^{-11}		
H_2NNH_2	1.2×10^{-6}		

3. J. N. Spencer, G. M. Bodner, and L. H. Rickard, *Chemistry: Structure & Dynamics*, Third Edition, John Wiley & Sons, 2006. Chapter 11: Problems: 95-98.

Model 2: Ions are Potential Acids or Bases.

All anions are *potential* bases:

$$Cl^-(aq) + H_2O(\ell) \rightleftarrows HCl(aq) + OH^-(aq)$$

$$NO_2^-(aq) + H_2O(\ell) \rightleftarrows HNO_2(aq) + OH^-(aq)$$

All cations are *potential* acids:

$$NH_4^+(aq) + H_2O(\ell) \rightleftarrows NH_3(aq) + H_3O^+(aq)$$

$$C_5H_5NH^+(aq) + H_2O(\ell) \rightleftarrows C_5H_5N(aq) + H_3O^+(aq)$$

Critical Thinking Questions

7. Which of the following ions are potential acids: Al^{3+} ; $CH_3NH_3^+$; HPO_4^{2-} ; F^- ; CH_3^- ?

8. Which of the following ions are potential bases: Fe^{3+} ; NH_4^+ ; F^- ?

Model 3: Some Potential Acids and Some Potential Bases Are so Weak That They Do Not Affect the pH of the Solution.

Any acid with a K_a less than 10^{-15} can be treated as if $K_a = 0$ (in water).

Any base with a K_b less than 10^{-15} can be treated as if $K_b = 0$ (in water).

The cations of alkali metals and alkaline earth metals act as neither acids nor bases in solution.

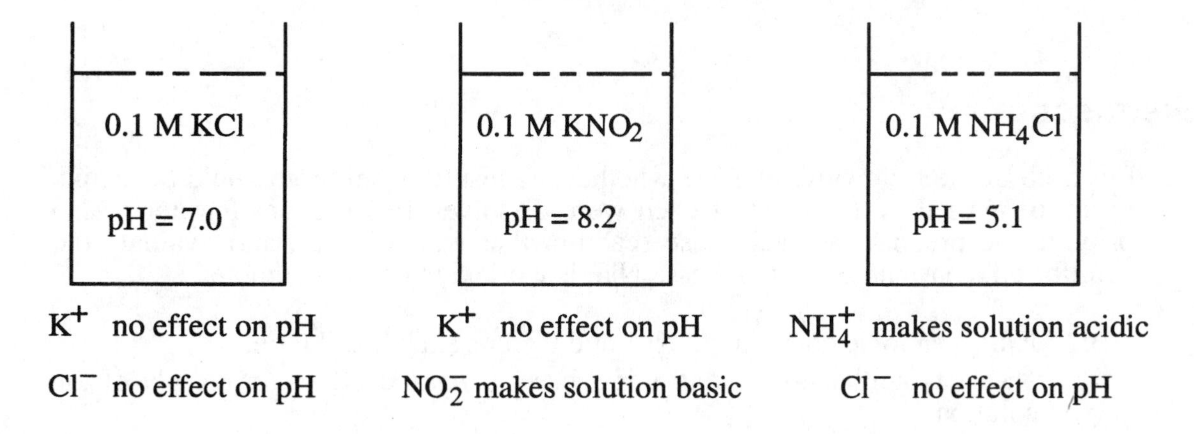

Table 1. The values of K_b for conjugate bases of some strong acids.

Acid	K_a	Conjugate Base	K_b
HBr	1×10^9		
HCl	1×10^6		
HNO_3	28		

Critical Thinking Questions

9. Fill in the missing entries in Table 1.

10. The chloride ion is the conjugate base of hydrochloric acid. Explain why Cl^- does not produce a basic solution when dissolved in water.

11. To 2 significant figures, predict the pH of a 1.00 molar solution of NaBr. Explain your reasoning carefully.

12. Explain why the conjugate bases of the strong acids do not produce basic solutions when dissolved in water.

Exercises

4. For each of the following, indicate whether the resulting solution would be acidic, basic, or neutral, if 1.0 mole of each were dissolved in 1.0 liter of water. Also provide the predominant acid-base reaction that would occur, and evaluate the equilibrium constant for that process. NaF is worked out as an example:

 i) NaF is an ionic compound. Na^+ and F^- ions exist in solution.
 ii) Na^+ acts neither as an acid nor a base; it has no effect on the pH of the solution.
 iii) The F^- ion is a weak base (the conjugate base of a weak acid). The solution will be basic.
 iv) The predominant reaction will be $F^-(aq) + H_2O \rightleftarrows HF(aq) + OH^-(aq)$. This is the chemical reaction that makes the solution basic.

 v) The equilibrium constant is $K_b = \dfrac{K_w}{K_a} = \dfrac{1.0 \times 10^{-14}}{7.0 \times 10^{-4}} = 1.4 \times 10^{-11}$

 Recall that the strong acids are: HCl; HBr; HI; $HClO_4$; H_2SO_4; HNO_3

 a) NH_4NO_3

 b) CsI

 c) CH_3COONa

 d) $KClO_4$

 e) magnesium(II) acetate

5. Determine the pH of each of the 1.0 M solutions in Ex. 4.

6. For each of the following, indicate whether the resulting solution would be acidic, basic, or neutral, if 1.0 mole of each were dissolved in 1.0 liter of water.

a) $NaCl$ b) KCl c) KNO_3

d) $NaCH_3CO_2$ (sodium acetate) e) NH_4Cl

f) NH_4NO_3 g) $NaNO_2$ h) $CaCl_2$

i) KCN j) KF k) $NaBr$

7. Given that K_a for HCN is 6.2×10^{-10}, calculate the pH of a 0.15 M KCN solution.

8. J. N. Spencer, G. M. Bodner, and L. H. Rickard, *Chemistry: Structure & Dynamics*, Third Edition, John Wiley & Sons, 2006. Chapter 11: Problems: 135, 146, 148.

Problem

1. One mole of $CH_3CH_2NH_3^+Cl^-$ is added to one liter of water. a) Will the solution be acidic, basic, or neutral? Explain briefly. b) Provide the chemical reaction that occurs to justify your answer in part a). That is, give the chemical reaction that causes the solution to be acidic or basic or neutral in accordance with your answer to part a).

ChemActivity 48

Redox Reactions

(Where Have All the Electrons Gone?)

Model 1: The Chemical Reaction of Zn(s) and Cu^{2+}(aq).

When a bar of zinc is placed in a 1.0 M copper(II) nitrate solution and left to stand for a while, solid copper is seen to deposit on the zinc bar, and some Zn^{2+} ions are found in solution. When equilibrium is reached in this system, essentially all of the copper ions have been plated out as solid copper (assuming that Cu^{2+} is the limiting reagent). Reactions such as this involve an explicit transfer of electrons between chemical species and are known as **oxidation-reduction**, or **redox**, reactions.

$$Zn(s) + Cu^{2+}(aq) \rightleftarrows Zn^{2+}(aq) + Cu(s) \quad (1)$$

gain lost

Critical Thinking Questions

1. Identify the reactant in equation (1) that:

   a) loses electrons Zn

   b) gains electrons Cu

2. How many electrons are transferred when:

   a) one Zn atom reacts with one Cu^{2+} ion?

   b) one mole of Zn reacts with one mole of Cu^{2+}?

3. Write the equilibrium expression, K, for reaction (1). Estimate [Zn^{2+}] and [Cu^{2+}] at equilibrium (according to the information given in Model 1). Comment on the magnitude of the equilibrium constant for this reaction.

Model 2: Oxidation and Reduction

Oxidation-reduction reactions are sometimes divided into half-reactions to separate and clarify the electron transfer process. The species that loses electrons is said to be oxidized, and the species that gains electrons is said to be reduced. The oxidized species is often referred to as the reducing agent. The substance that is reduced is referred to as the oxidizing agent.

The half-reactions that describe the electron transfer process are:

$$Zn(s) \rightleftarrows Zn^{2+}(aq) + 2e^-$$

$$Cu^{2+}(aq) + 2e^- \rightleftarrows Cu(s)$$

4. Which species is oxidized in equation (1)? Reduced?

5. Which species is the oxidizing agent in equation (1)? The reducing agent?

6. How can the electron transfer process be stopped once the zinc has been placed into the Cu^{2+} solution?

Model 3: Results of Placing Metal Bars in a Variety of Solutions at 25 °C.

Metal Bar	Ion Solution (1.0 M)	Concentration of Metal Ions at Equilibrium, M		K
Zn	Cu^{2+}	$[Cu^{2+}] \approx 0$	$[Zn^{2+}] \approx 1.0$	
Zn	K^{+}	$[K^{+}] \approx 1.0$	$[Zn^{2+}] \approx 0$	
Co	Ni^{2+}	$[Ni^{2+}] \approx 0.1$	$[Co^{2+}] \approx 0.9$	
Co	Cu^{2+}	$[Cu^{2+}] \approx 0$	$[Co^{2+}] \approx 1.0$	
Co	Cr^{3+}	$[Cr^{3+}] \approx 1.0$	$[Co^{2+}] \approx 0$	

The results were obtained with metal bars large enough so that the limiting reagent in any redox reaction with the solution was the ion in solution.

Critical Thinking Questions

7. For each of the five experiments described in Model 3, write the balanced chemical equation (no "e^-" appears in the balanced chemical equation) for the redox reaction that **could** occur between the metal bar and the ion in solution. Note that the same number of electrons must be lost and gained in the transfer process. In each case indicate the oxidizing agent and the reducing agent.

8. Fill in the "K" column in Model 3 by indicating whether K is >1, <1, or impossible to deduce from the data given.

9. Based on the data in the first two rows of Model 3, which do you think would be considered a stronger oxidizing agent, Cu^{2+} or K^{+}?

10. Based on the data in the last three rows of Model 3, rank the strength as oxidizing agents of the metal ions Ni^{2+}, Cu^{2+}, and Cr^{3+}.

11. If possible, rank the metal ions Ni^{2+}, Cu^{2+}, Cr^{3+}, and K^{+} in terms of their strength as oxidizing agents. If this is not possible, rank as many as you can and propose an experiment (or series of experiments) that would enable you to complete the rankings.

Exercises

1. Identify the reducing agent and the oxidizing agent in each of the following reactions. All of these reactions have $K > 1$.

   a) $Br_2(aq) + Hg(s) \rightleftarrows 2\,Br^-(aq) + Hg^{2+}(aq)$
   b) $2\,Co^{3+}(aq) + 2\,Br^-(aq) \rightleftarrows Br_2(aq) + 2\,Co^{2+}(aq)$
   c) $Cl_2(aq) + 2\,Br^-(aq) \rightleftarrows 2\,Cl^-(aq) + Br_2(aq)$
   d) $2\,H^+(aq) + Zn(s) \rightleftarrows H_2(aq) + Zn^{2+}(aq)$
   e) $S_2O_8^{2-}(aq) + Zn(s) \rightleftarrows Zn^{2+}(aq) + 2\,SO_4^{2-}(aq)$
   f) $Au^{3+}(aq) + Fe(s) \rightleftarrows Au(s) + Fe^{3+}(aq)$

2. Assume that all of the stoichiometric coefficients for the reactions in Ex. 1 represent molar quantities. How many electrons are transferred when each reaction takes place?

3. Indicate whether the following statement is true or false and explain your reasoning.

   $Cu^{2+}(aq)$ is a stronger oxidizing agent than $K^+(aq)$.

Problem

1. Describe an experiment that would allow you to determine the relative strengths of zinc and nickel metals as reducing agents. Provide enough detail so that another student in your class could understand what to do, and also indicate what the observed results of the experiment would be. Make sure that you also indicate which of the two metals IS the stronger reducing agent.

ChemActivity **49**

Oxidation Numbers

Information

Oxidation numbers are an accounting system for electrons (Lewis structures and formal charge are also accounting systems for electrons). One of the main uses of oxidation numbers (but not the only use) is to locate the oxidized and reduced species in redox reactions. For example, the oxidized and reduced species are not obvious in the following reactions:

$$5\,Cr^{3+}(aq) + 3\,MnO_4^{-}(aq) + 8\,H_2O \rightleftarrows 5\,CrO_4^{2-}(aq) + 3\,Mn^{2+}(aq) + 16\,H^{+}(aq) \qquad (1)$$

$$2\,CuI(s) \rightleftarrows Cu(s) + Cu^{2+}(aq) + 2\,I^{-}(aq) \qquad (2)$$

In an oxidation-reduction reaction, the species that is oxidized undergoes an increase in oxidation number, and the species that is reduced undergoes a decrease in oxidation number.

For pure ionic substances, the oxidation number and the charge on the ion are often the same. In NaCl, for example, the oxidation number and the charge on the sodium is +1, and the oxidation number and the charge on the chlorine is –1. It is important to realize, however, that for covalent polar and nonpolar molecules, oxidation numbers have little relationship to the actual charges on the atoms within the molecule. In CH_4, for example, the oxidation number on the carbon is –4 and the oxidation number of each hydrogen is +1. We know, however, that the partial charge on the carbon atom is much closer to zero than it is to –4 because the difference in the electronegativities of carbon and hydrogen is small. (A MOPAC calculation yields: $\delta_C = -0.266$; $\delta_H = +0.066$.)

Model: Oxidation Number Conventions.

Oxidation numbers are often written above the atomic symbol.

$$\overset{+1}{H}\overset{+7}{Cl}\overset{-2}{O_4} \qquad \overset{0}{O_2} \qquad \overset{-4}{C}\overset{+1}{H_4}$$

Oxidation Numbers

- The oxidation number is 0 in any neutral substance that contains atoms of only one element. Aluminum foil, iron metal, and the H_2, O_2, O_3, P_4, and S_8 molecules all contain atoms that have an oxidation number of 0.
- The oxidation number is equal to the charge on the ion for ions that contain only a single atom. The oxidation number of the Na^{+1} ion, for example, is +1, whereas the oxidation number of Cl^{-} is –1.
- The oxidation number of hydrogen is +1 when it is combined with a *more electronegative element*. The oxidation number of hydrogen is +1 in the CH_4, NH_3, H_2O, and HCl molecules.

- The oxidation number of hydrogen is –1 when it is combined with a *less electronegative element*. The oxidation number of hydrogen is –1 in the LiH, NaH, CaH_2, and $LiAlH_4$ molecules.
- The elements of Groups IA and IIA form compounds in which the metal atoms have oxidation numbers of +1 and +2, respectively.
- Oxygen usually has an oxidation number of –2. Exceptions include molecules and polyatomic ions that contain O–O bonds: O_2, O_3, H_2O_2, and the O_2^{2-} ion.
- Elements in Group VIIA have an oxidation number of –1 when the atom is bonded to a less electronegative element. The oxidation number of each chlorine atom in CCl_4 is –1.
- The sum of the oxidation numbers of the atoms in a neutral substance is zero.
  H_2O (2 hydrogen)(+1) + (1 oxygen)(–2) = 0
- The sum of the oxidation numbers of the atoms in a polyatomic ion is equal to the charge on the ion.
  OH^- (1 hydrogen)(+1) + (1 oxygen)(–2) = –1
- The least electronegative element is assigned a positive oxidation state. Sulfur is assigned a positive oxidation state in SO_2 because it is less electronegative than oxygen.
  SO_2 (1 sulfur)(+4) + (2 oxygen)(–2) = 0

Oxidation Numbers for Organic Molecules

To assign oxidation numbers to atoms in organic molecules, we *treat* each bond as if it were an ionic bond and the electrons belonged to the more electronegative element. Then,

$$OX_a = G_a - V_a$$

where OX_a is the oxidation number of atom "a", G_a is the group number of the atom, and V_a is the number of valence electrons assigned to the atom in the Lewis structure of the molecule.

Acetic acid is shown as an example:

Write Lewis structure

```
      H
      |     ••
   H— C — C — O — H
      |   ||  ••
      H  :O:
```

Assign electrons ionically

```
      H
      ••              ••
   H :C•    •C       :O:    H
      ••              ••
             ••
      H     :O:
```

Calculate oxidation numbers

methyl C atom	OX = 4 – 7 = –3
carbonyl C atom	OX = 4 – 1 = +3
each H atom	OX = 1 – 0 = +1
each oxygen atom	OX = 6 – 8 = –2

Critical Thinking Questions

1. Which is the better representation of the actual charge on an atom in a molecule—the formal charge; the partial charge; the oxidation state? Why?

2. In which of the following are the oxidation number and the partial charge the same number? MgO ; CO_2 ; NaF ; H_2O ; CCl_4 ; $NiCl_2$.

3. Find the oxidation numbers of all atoms in equation (1)—on the left-hand side and on the right-hand side of the equation. Does any atom increase its oxidation number?

4. Is a species that has an atom for which there is an increase in oxidation number oxidized or reduced? Is that species the oxidizing agent or the reducing agent?

5. Which species in equation (1) is oxidized? Which species is reduced?

6. Which species in equation (2) is oxidized? Which species is reduced?

7. Describe, in grammatically correct English sentences, how one can determine whether or not a reaction is an oxidation-reduction reaction.

Exercises

1. Give the oxidation number for each atom in the following molecules: Br_2 ; $NaCl$; $CuCl_2$; CH_4 ; CO_2 ; $SiCl_4$; CCl_4 ; SCl_2 ; Br_2O.

2. Give the oxidation number for each atom in the following ions: Ni^{2+} ; NO_3^- ; CO_3^{2-}; SO_4^{2-} ; NH_4^+ ; ClO_4^- ; MnO_4^- ; CN^- ; IF_4^+ ; PO_4^{3-} .

3. Give the oxidation number for each atom in the following molecules: $NiCl_2$; HNO_3 ; Na_2CO_3 ; $Al_2(SO_4)_3$; NH_4Cl ; $KMnO_4$; KCN ; $HClO_4$.

4. Give the oxidation number for each atom in the following ions: HCO_3^- ; HSO_4^- ; $H_2PO_4^-$; NH_2^- ; $Cr_2O_7^{2-}$.

5. Give the oxidation number for each atom in the following molecules: CH_3OH ; CH_3CH_2OH ; H_2CCH_2 ; CH_3Cl ; CCl_4 .

6. Give the oxidation number of N and H in NH_3. What is the oxidation number of Cu in $Cu(NH_3)_4^{2+}$?

7. Give the oxidation number of O and H in OH^-. What is the oxidation number of Al in $Al(OH)_4^-$?

8. An oxidation number need not be an integer. Give the oxidation number for each atom in the following molecules: P_4O_7 ; P_4O_6 ; P_4O_8 ; P_4O_9 .

9. Which of the following are redox reactions?
   a) $3\,H_2(g) + N_2(g) \rightleftarrows 2\,NH_3(g)$
   b) $Ag^+(aq) + Cl^-(aq) \rightleftarrows AgCl(s)$
   c) $C(s) + O_2(g) \rightleftarrows CO_2(g)$
   d) $H_2CCH_2(g) + H_2(g) \rightleftarrows H_3CCH_3(g)$
   e) $3\,Cu(s) + 8\,H^+(aq) + 2\,NO_3^-(aq) \rightleftarrows 3\,Cu^{2+}(aq) + 2\,NO(g) + 4\,H_2O$
   f) $H_2(g) + Cl_2(g) \rightleftarrows 2\,HCl(g)$
   g) $Cu^{2+}(aq) + 4\,NH_3(aq) \rightleftarrows Cu(NH_3)_4^{2+}(aq)$

10. When natural gas (methane) burns, the chemical reaction is

$$CH_4(g) + 2\,O_2(g) \rightleftarrows CO_2(g) + 2\,H_2O(g) \quad .$$

Is this an oxidation-reduction reaction?

11. When iron corrodes, the chemical reaction is

$$2\,Fe(s) + O_2(aq) + 2\,H_2O(l) \rightleftarrows 2\,FeO{\bullet}H_2O(s) \quad .$$

Is this an oxidation-reduction reaction?

12. Plants convert carbon dioxide and water into carbohydrates and dioxygen by a series of reactions called photosynthesis. The overall chemical reaction is

$$6\,CO_2(g) + 6\,H_2O(l) \rightleftarrows C_6H_{12}O_6(aq) + 6\,O_2(g)$$

Is this an oxidation-reduction reaction?

13. J. N. Spencer, G. M. Bodner, and L. H. Rickard, *Chemistry: Structure & Dynamics*, Third Edition, John Wiley & Sons, 2006. Chapter 12: Problems: 8, 10, 12-15.

Problem

1. Give the oxidation number for the bromine atom in each of the species below, and then describe the relationship between the oxidation number on the bromine and the relative acidity of these compounds: $HOBrO_2$; $HOBr$; $HOBrO$.

ChemActivity 50

The Electrochemical Cell

(How Does a Battery Work?)

Model 1: Schematic of a Galvanic Cell.

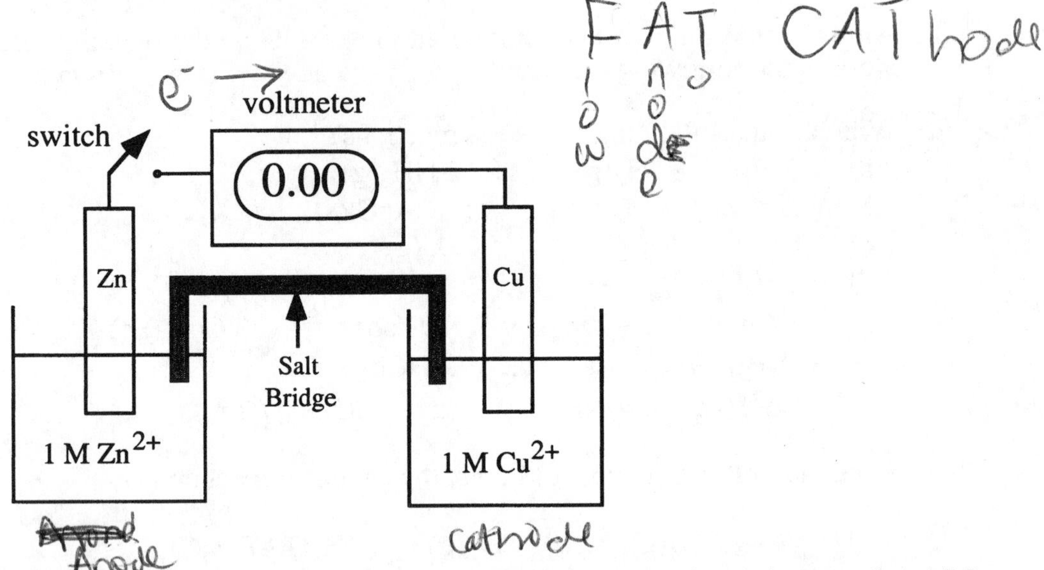

The beaker on the left contains 1 M $Zn(NO_3)_2$ and the beaker on the right contains 1 M $Cu(NO_3)_2$.

It is possible to design a redox reaction such that the oxidation occurs at one location and the reduction occurs at another location. The device is called a **galvanic** or **voltaic** cell. The **cathode** (usually a metal bar or carbon rod) is the electrode where reduction takes place; the **anode** (usually a metal bar or carbon rod) is the electrode where oxidation takes place. The salt bridge allows ions to slowly migrate from one beaker to the other to maintain electrical neutrality in each half-cell. The voltmeter measures the voltage (or potential), V, between the two electrodes. If the temperature is 298 K, and the solutions are 1 M, then the beakers with the electrodes are each considered to be a **standard half-cell**.

When the switch is closed, the following is observed in the model:

- The mass of the Cu electrode increases and the concentration of $Cu^{2+}(aq)$ decreases.
- The mass of the Zn electrode decreases and the concentration of $Zn^{2+}(aq)$ increases.
- Electrons (e^-) are observed to flow through the wire.
- The voltage measured with the voltmeter is 1.10 V.

Critical Thinking Questions

1. What is the half-reaction occurring in the copper half-cell?

   $Zn \rightarrow Zn^{2+} + 2e^-$

   $Cu^{2+} + 2e^- \rightarrow Cu$

2. What is the half-reaction occurring in the zinc half-cell?

   $Zn + Cu^{2+} \rightarrow Zn^{2+} + Cu$

3. What is the overall (net) chemical reaction taking place in the galvanic cell? (Note: This reaction should not have any e^- in it.)

4. Label the anode and the cathode in the diagram.

5. In which direction (through the wire) are the electrons flowing?

6. Electrons flow from the negative electrode to the positive electrode. Which electrode, Zn or Cu, is the negative electrode?

   (−) (+)

7. For the cell in the model:

   a) What species is oxidized in the overall process? Zn

   b) What species is reduced in the overall process? Cu

8. Is there any way to stop the electron transfer process once the switch has been closed?

   break the salt bridge

9. What use(s) could be made of the flow of electrons in the wire?

   energy

10. Give two advantages of a voltaic cell, as described in the model, compared to inserting a zinc bar into a Cu^{2+} solution.

    last longer

    continues the cycle

Model 2: The Standard Hydrogen Electrode.

The chemical processes taking place in a galvanic cell may be viewed as a "tug-of-war" for electrons between the two half-cells. The "winner" is the one containing the stronger oxidizing agent—it is the one that gains the electrons and gets reduced. The voltage is a measure of the difference in electron-pulling strength.

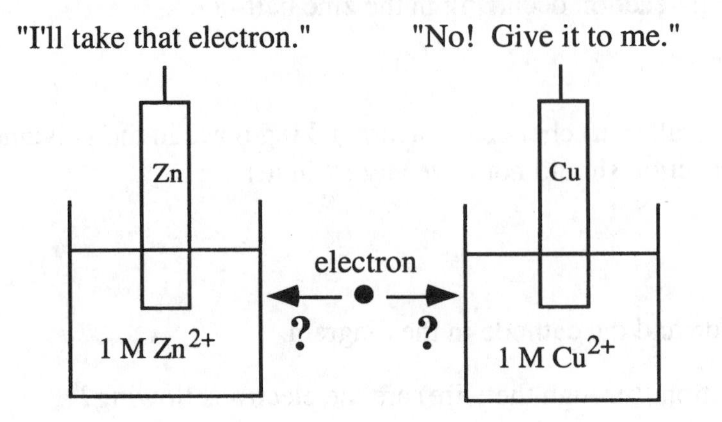

As a standard basis for comparing relative electron-pulling strength (also referred to as **reduction potential**), the Standard Hydrogen Electrode (SHE) is often used. This half-cell consists of a platinum electrode (Pt is chemically inert, but it is an excellent conductor of electricity) submerged in a 1 M solution of H^+ ions (this designation is used rather than H_3O^+, but the meaning is the same) at 298 K, and bathed by H_2 gas at 1 atm pressure.

Figure 1. The standard hydrogen electrode.

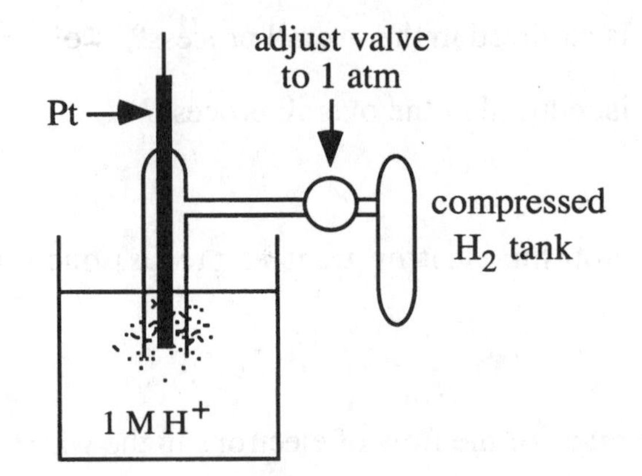

The **standard reduction potential** (reduction potential under standard conditions), $E°_{Red}$, of the SHE is defined as zero volts.

$$2e^- + 2H^+(1\text{ M}) \rightleftarrows H_2(g;1\text{ atm}) \qquad E°_{Red} \equiv 0.00\text{ V}$$

When a SHE is connected to the Cu/Cu^{2+} half-cell from the model, the Cu/Cu^{2+} half-cell exhibits a stronger pull on electrons than does the SHE half-cell. Thus, the following reaction takes place at the Cu electrode (cathode):

$$Cu^{2+}(1\ M) + 2e^- \rightleftarrows Cu(s)$$

Simultaneously, at the Pt electrode (anode), the following reaction takes place:

$$H_2(g;\ 1\ atm) \rightleftarrows 2H^+(1\ M) + 2e^-$$

The experimental voltage, $E°$, is 0.34 V.

Critical Thinking Questions

11. Which is the stronger oxidizing agent, $Cu^{2+}(aq)$ or $H^+(aq)$?

12. In terms of volts, how much stronger is the stronger of the two oxidizing agents in CTQ 11?

13. What value (in volts) should be assigned as the standard reduction potential, $E°_{Red}$, of the Cu/Cu^{2+} half-cell?

Exercises

1. For the cell in the model, which is the stronger oxidizing agent—Zn^{2+} or Cu^{2+}? Cu

2. For the cell in the model, how much stronger (in terms of volts) is the stronger oxidizing agent?

3. Draw a galvanic cell consisting of a SHE and the copper electrode described above. Indicate a) the anode and the cathode, b) the direction of flow of the electrons in the wire, and c) which electrode is positive and which electrode is negative. Write down the half-reactions that are occurring at each electrode, and then write down the overall chemical process occurring in the cell.

4. When a standard $Al(s)/Al^{3+}$ cell is connected to a SHE, the electrons are observed to flow in the direction of the SHE. The voltage is measured as 1.66 V.
   a) Sketch this electrochemical cell.
   b) Identify the anode and the cathode in this system.
   c) Identify the positive and negative electrode.
   d) Give the half-reaction occurring in each half-cell, and then give the net chemical reaction for the cell. Keep in mind that the number of electrons being given up and being received must be the same.
   e) What is the standard reduction potential for the $Al(s)/Al^{3+}$ half-cell? Explain your reasoning.

5. J. N. Spencer, G. M. Bodner, and L. H. Rickard, *Chemistry: Structure & Dynamics*, Third Edition, John Wiley & Sons, 2006. Chapter 12: Problems: 27, 33.

ChemActivity 51

The Cell Voltage

Model 1: The Cell Voltage.

The Cu/Cu^{2+} half-reaction can be written as an oxidation or as a reduction. When a half-reaction is reversed, the magnitude of the driving force, the voltage, remains the same but the sign is reversed.

$$Cu^{2+}(1\text{ M}) + 2e^- \rightleftarrows Cu(s) \qquad E^\circ_{Red} = +0.34\text{ V}$$
$$Cu(s) \rightleftarrows Cu^{2+}(1\text{ M}) + 2e^- \qquad E^\circ_{Ox} = -0.34\text{ V}$$

We recognize that every redox reaction consists of an oxidation and a reduction. And the overall driving force or voltage, E, will be a sum of the driving forces for the reduction and the oxidation reaction.

$$E = E_{Ox} + E_{Red}$$

When the reaction takes place under standard conditions (soluble species at 1 M and gases at 1 atm):

$$E^\circ = E^\circ_{Ox} + E^\circ_{red}$$

Example 1: A Cu/Cu^{2+} half-cell and a SHE

$$Cu^{2+}(1\text{ M}) + 2e^- \rightleftarrows Cu(s) \qquad E^\circ_{Red} = 0.34\text{ V}$$
$$H_2(g;\ 1\text{ atm}) \rightleftarrows 2e^- + 2H^+(1\text{ M}) \qquad E^\circ_{Ox} = 0.00\text{ V}$$

$$Cu^{2+}(1\text{ M}) + H_2(g;\ 1\text{ atm}) \rightleftarrows Cu(s) + 2H^+(1\text{ M}) \qquad E^\circ = 0.34\text{ V}$$

Example 2: A Zn/Zn^{2+} half-cell and a SHE

$$Zn(s) \rightleftarrows Zn^{2+}(1\text{ M}) + 2e^- \qquad E^\circ_{Ox} = 0.76\text{ V}$$
$$2H^+(1\text{ M}) + 2e^- \rightleftarrows H_2(g;\ 1\text{ atm}) \qquad E^\circ_{Red} = 0.00\text{ V}$$

$$Zn(s) + 2H^+(1\text{ M}) \rightleftarrows Zn^{2+}(1\text{ M}) + H_2(g;\ 1\text{ atm}) \qquad E^\circ = 0.76\text{ V}$$

Critical Thinking Questions

1. Consider a galvanic cell composed of a Cu/Cu^{2+}(1 M) half-cell and a Zn/Zn^{2+} (1 M) half cell.

   a) Which electrode has the stronger pull on electrons, Cu/Cu^{2+}(1 M) or Zn/Zn^{2+}(1 M)?

   b) Write the cathode reaction.

   $Cu^{2+} + 2e^- \rightarrow Cu$

   c) Write the anode reaction.

   $Zn \rightarrow Zn^{2+} + 2e^-$

   d) Do the electrons flow from the cathode to the anode or from the anode to the cathode?

   anode → Cathode

   e) Use the standard potentials found above to confirm that the cell voltage, $E°$, is 1.10 V.

   Cu .34
   Zn +.76
   1.10V

Model 2: Measured Voltages for Some Galvanic Cells Using Standard Electrodes (all ions and soluble species at 1 M and all gases at 1 atm).

Cathode	Anode	$E°$ (V)
Cu/Cu^{2+}	Zn/Zn^{2+}	1.10
Cu/Cu^{2+}	SHE	0.34
Br_2/Br^-	Zn/Zn^{2+}	1.85
Zn/Zn^{2+}	K/K^+	2.16
Cl_2/Cl^-	Ag/Ag^+	0.56
Ag/Ag^+	K/K^+	3.72

Critical Thinking Questions

2. Determine the standard reduction potential, $E°_{Red}$, for each of the following half-reactions:

$Cl_2 + 2e^- \rightleftarrows 2\,Cl^-$

$Br_2 + 2e^- \rightleftarrows 2\,Br^-$

$Ag^+ + e^- \rightleftarrows Ag$

$Cu^{2+} + 2e^- \rightleftarrows Cu$

$2H^+ + 2e^- \rightleftarrows H_2$

$Zn^{2+} + 2e^- \rightleftarrows Zn$

$K^+ + e^- \rightleftarrows K$

3. Examine the results from CTQ 2.

   a) What is the strongest oxidizing agent?

   b) What is the weakest oxidizing agent?

4. The stronger the oxidizing agent, the weaker the resulting reducing agent that is produced by the acquisition of electrons. In this case:

   a) What is the strongest reducing agent on the right-hand side of the list in CTQ 2?

   b) What is the weakest reducing agent?

Exercises

Use a table of standard reduction potentials for the following exercises.

1. You decide to construct a zinc/aluminum galvanic cell in which the electrodes are connected by a wire and the solutions are connected with a salt bridge. One electrode consists of an aluminum bar in a 1.0 M solution of aluminum(III) nitrate. The other electrode consists of a zinc bar in a 1.0 M solution of zinc(II) nitrate. Zinc(II) has a more positive standard reduction potential than Al(III).
   a) Which electrode is the cathode and which is the anode?
   b) What is the direction of electron flow?
   c) Which electrode is negative? Positive?
   d) What chemical reactions are occurring at each electrode?
   e) What is the overall chemical reaction?
   f) After a period of time, will the bar of zinc become heavier, lighter, or stay the same weight? Will the bar of aluminum become heavier, lighter, or stay the same weight?

2. Indicate whether each of the following statements is true or false and explain your reasoning:
   a) The half-cell with the larger standard reduction potential is always the anode in a galvanic cell.
   b) Whenever an oxidation half-reaction takes place, a reduction half-reaction must also take place.

3. For each of the chemical reactions below assume that at the start of the reaction the concentration of each ion (on the left-hand side and on the right-hand side) is 1.0 M, each gas (on the left-hand side and on the right-hand side) has a pressure of 1.0 atm, and each solid (on the left-hand side and on the right-hand side) is present. Determine the cell voltage for each reaction, as written. Which of these reactions will proceed to the right? Which of these reaction will proceed to the left?

   a) $Cr(s) + Pb^{2+}(aq) \leftrightarrows Pb(s) + Cr^{2+}(aq)$

   b) $H_2(g) + 2Ag^{+}(aq) \leftrightarrows 2\,Ag + 2H^{+}(aq)$

   c) $2Cr^{2+}(aq) + Mg^{2+} \leftrightarrows 2Cr^{3+}(aq) + Mg(s)$

   d) $NO_2^{-}(aq) + ClO^{-}(aq) \leftrightarrows NO_3^{-}(aq) + Cl^{-}(aq)$

   e) $4AgBr(s) + 4OH^{-}(aq) \leftrightarrows O_2(g) + 2H_2O + 4Ag(s) + 4Br^{-}(aq)$

4. For each of the chemical reactions in Exercise 3 assume that only the reactants (the species on the left-hand side) are present. Which of these chemical reactions will occur? Which will not occur?

5. J. N. Spencer, G. M. Bodner, and L. H. Rickard, *Chemistry: Structure & Dynamics*, Third Edition, John Wiley & Sons, 2006. Chapter 12: Problems: 42-44, 50, 52, 54, 57, 65, 68, 70, 73, 74, 131, 136, 142, 144.

Problems

1. Assuming standard conditions, indicate which of the following is true:

   a) $H_2(g)$ can reduce $Ag^+(aq)$
   b) $H_2(g)$ can reduce $Ni^{2+}(aq)$
   c) $Fe^{2+}(aq)$ can reduce $Cu^{2+}(aq)$
   d) $H^+(aq)$ can oxidize $Mg(s)$
   e) $Pb^{2+}(aq)$ can oxidize $Ni(s)$

2. A student places some Zn(s) powder in a beaker of 1 M nitric acid, and some Cu(s) powder in another beaker also containing 1 M nitric acid. In which, if any, of the beakers would you expect the solid to react and evolve hydrogen gas? Explain your reasoning.

3. Find a reagent that can oxidize Br^- to Br_2 but cannot oxidize Cl^- to Cl_2.

4. Use a table of standard reduction potentials to determine whether or not a reaction occurs when a tin (Sn) rod is placed into 500 mL of 1.0 M Ni^{2+}. Briefly explain.

5. A 20.00 mL sample of oxalic acid solution, $H_2C_2O_4$, was titrated with 0.256 M $KMnO_4$ solution. What is the molarity of the oxalic acid solution if it took 14.6 mL of the $KMnO_4$ solution to completely react with the oxalic acid? The oxidation-reduction reaction is:

$$5\ H_2C_2O_4\ (aq) + 2\ MnO_4^-(aq) + 6\ H^+(aq) \rightleftarrows 10\ CO_2(g) + 2\ Mn^{2+}(aq) + 8\ H_2O(\ell)$$

ChemActivity 52

Entropy (I)

(Why Is my Desk so Messy?)

Water flows downhill, not uphill. Ice melts on a warm day; water does not freeze on a warm day. We eat an apple, and we excrete CO_2 and H_2O. If we ingest CO_2 and H_2O, we do not expect to excrete an apple. We would be amazed if a ten-year-old ostrich gradually became younger and, ten years later, had become yolk and albumen encased in a shell. If we throw a handful of confetti out of a window, we do not expect all of the confetti to accumulate in the refuse container at the end of the street.

Similarly, when a piece of zinc metal dissolves in a strong acid solution, bubbles of hydrogen gas evolve.

$$Zn(s) + 2\,H_3O^+(aq) \rightleftarrows Zn^{2+}(aq) + H_2(g) + 2\,H_2O(l) \qquad (1)$$

Although perhaps not as familiar as the melting of ice, this process is also not surprising. However, if we saw a video in which H_2 bubbles formed at the surface of a solution and sank through the solution until they disappeared, while a strip of zinc metal formed in the middle of the solution, we would likely think that the video was being run backward.[1]

Most important of all, my desk gets messy.

Clearly, many physical and chemical processes proceed **naturally** in one direction, but not in the other. (They are sometimes referred to as being **spontaneous** in the direction in which they proceed naturally.) This raises the question: What factor (or factors) determine the direction in which reactions proceed naturally?

Model 1: A Ball Tends to Roll Downhill.

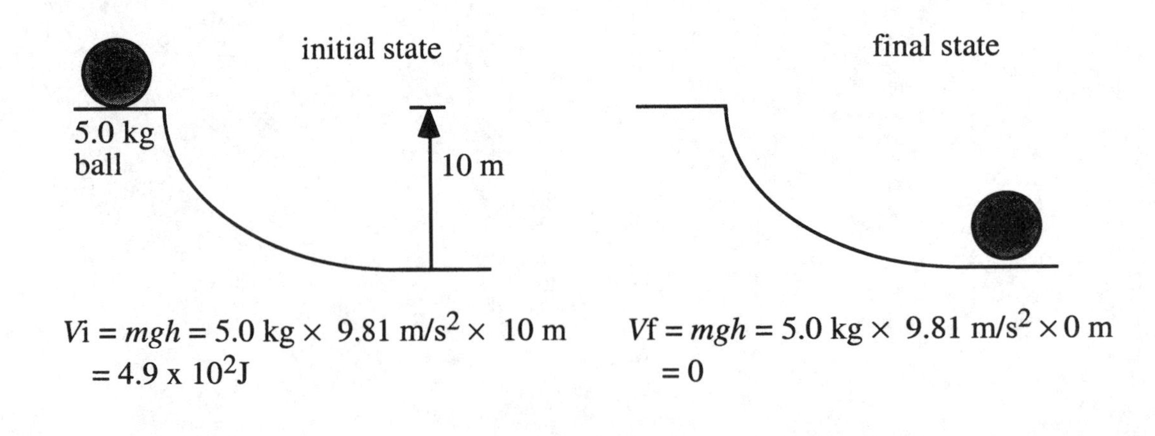

[1] J. N. Spencer, G. M. Bodner, and L. H. Rickard, *Chemistry: Structure & Dynamics*, Third Edition, John Wiley & Sons, 2006, Section 13.1.

Critical Thinking Questions

1. A ball tends to roll down a hill.

   a) Which is the lower energy state: the ball at the top of the hill or the ball at the bottom of the hill?

   b) Is the change in the potential energy for this process, $V_f - V_i$, positive, negative, or zero?

   c) Why doesn't the ball roll up the hill?

Model 2: The Formation of Solid Sodium Chloride From Gaseous Ions.

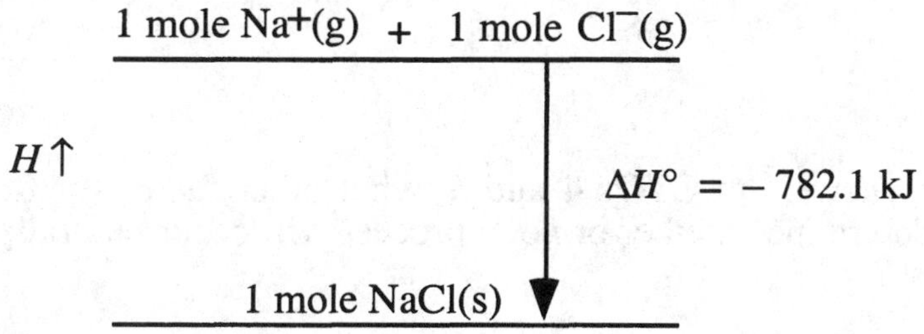

1 mole $Na^+(g)$ + 1 mole $Cl^-(g)$

$H\uparrow$ $\Delta H° = -782.1$ kJ

1 mole NaCl(s)

Critical Thinking Questions

2. Gaseous sodium ions and gaseous chloride ions will combine to form solid sodium chloride.

   a) Which is the lower energy state: 1 mole of NaCl(s) or 1 mole of $Na^+(g)$ and 1 mole of $Cl^-(g)$?

   b) Is ΔH for the process $Na^+(g) + Cl^-(g) \rightleftarrows NaCl(s)$ positive, negative, or zero?

   c) Why doesn't a salt crystal suddenly become gaseous sodium ions and gaseous chloride ions?

3. Write a chemical equation that describes the melting of ice, and indicate whether ΔH for the process is positive, negative, or zero. Under what temperature conditions does this process naturally occur?

4. Write a chemical equation that describes the freezing of water, and indicate whether ΔH for the process is positive, negative, or zero. Under what temperature conditions does this process naturally occur?

5. Is it possible to determine whether a process will occur naturally solely by examining the sign of ΔH for the process? Explain.

6. Based on your answers to CTQs 4 and 5, what other factor (or factors) must be considered to determine whether or not a process will occur naturally under a given set of conditions?

Information

Many naturally occurring processes tend to be exothermic, but this is not a requirement. The temperature at which a process occurs also plays a role in determining in which direction a process will proceed naturally. Thus there appear to be two important factors in this determination—the ΔH of the reaction and another factor whose impact is influenced by the temperature. This second factor, known as **entropy**, ***S***, is a measure of disorder or randomness. The more disorder, the higher (more positive) the entropy. The entropy can never be less than zero; that is, entropy values are always positive.

Model 3: Gaseous Molecules in a Box.

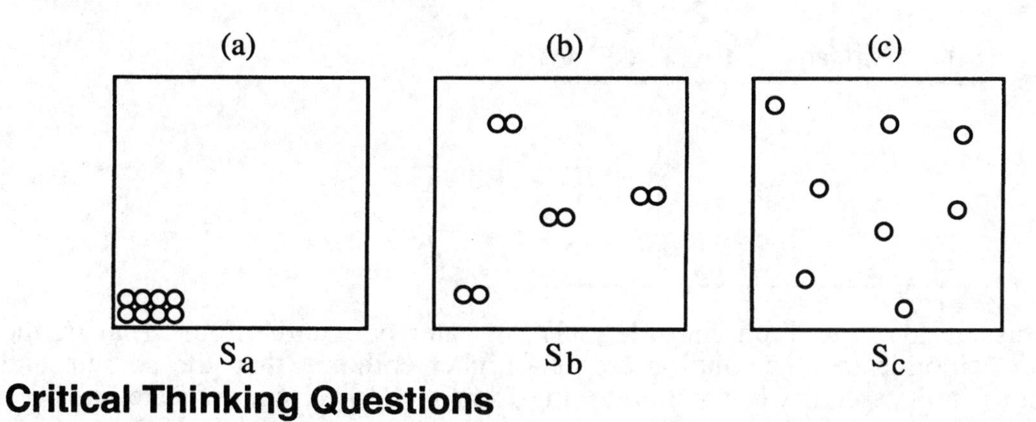

Critical Thinking Questions

7. In which case in Model 3 are the molecules most disordered: a), b), or c)?

8. Which case in Model 3 has the greatest entropy: a), b), or c)?

9. Suppose that at some temperature the naturally occurring process starts with case a) and ends with case c). Is $S_a > S_c$ or is $S_c > S_a$? Is ΔS for this naturally occurring process positive or negative?

10. Suppose that at some temperature (different than the temperature in CTQ 9) the naturally occurring process starts with case c) and ends with case a). Is ΔS for this naturally occurring process positive or negative?

Model 4: Sodium Chloride Does Dissolve in Water!

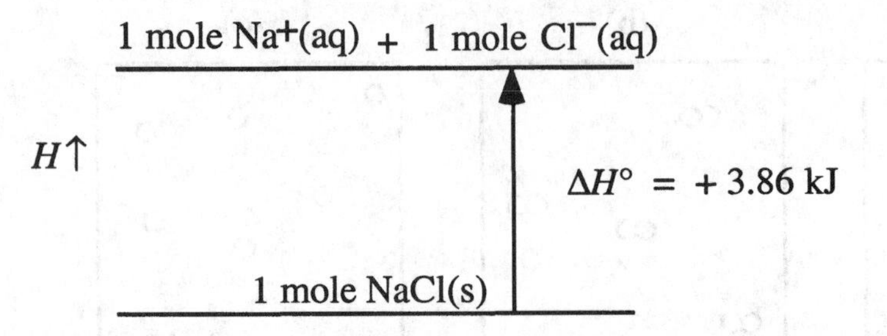

The dissolution of sodium chloride (salt) in water is endothermic. That is, the sodium and chloride ions in solution are at a higher enthalpy than the sodium and chloride ions in the solid. Clearly, the driving force to a lower energy state has been overcome in this case. Another factor that determines the direction in which reactions proceed naturally is entropy: chemical (and physical) processes tend to be driven toward the state of highest entropy (to most disorder).

The following are some generalizations that can often be used in considering the entropy associated with a chemical species, or the change in entropy (ΔS) associated with a chemical reaction:

- As the number of particles in the system increases, the amount of disorder increases (ΔS is positive).
- As the volume in which particles can move increases, the amount of disorder increases (ΔS is positive).
- As the temperature of a system increases, the motion of the particles and the amount of disorder increases (ΔS is positive).

Critical Thinking Questions

11. Is ΔS for the following process positive or negative? Why?

$C_4H_8(g) \rightleftarrows 2\, C_2H_4(g)$

12. Is ΔS for the following process positive or negative? Why?

C_4H_8(g; 298 K; 1 atm) $\rightleftarrows$ C_4H_8(g; 298 K; 0.5 atm)

13. Is ΔS for the following process positive or negative? Why?

    $C_4H_8(g;\ 298\ K;\ 1\ atm) \rightleftarrows C_4H_8(g;\ 398\ K;\ 1\ atm)$

14. Rank the following in order of increasing entropy: $H_2O(g)$; $H_2O(l)$; $H_2O(s)$. In general, how are the entropy of the solid, liquid, and gaseous phases of a particular compound related? Explain your reasoning.

Exercises

1. For each of the following processes, predict whether $\Delta S°$ (for the chemical reaction) is expected to be positive or negative. Explain your reasoning.

   a) $N_2(g) + 3H_2(g) \rightleftarrows 2\ NH_3(g)$
   b) $CO_2(g) \rightleftarrows CO_2(s)$
   c) $CaCO_3(s) \rightleftarrows CaO(s) + CO_2(g)$
   d) The air in a balloon escapes out a hole and the balloon flies wildly around the room. (Consider ΔS for the air molecules originally in the balloon.)
   e) A precipitate of $Pb(OH)_2$ forms when solutions of lead(II) nitrate and sodium(I) hydroxide are mixed.

2. Consider the reactions:

   $$N_2(g) + O_2(g) \rightleftarrows 2\ NO(g)$$
   $$N_2(g) + 2\ O_2(g) \rightleftarrows N_2O_4(g)$$

   How would you expect the values of ΔS for these reactions to compare? That is, would they be equal, and if not, which one would be larger? Explain your reasoning.

3. Indicate whether the following statement is true or false and explain your reasoning.

   For the reaction $2\ SO_3(g) \leftrightarrows 2\ SO_2(g) + O_2(g)$, $\Delta S°$ is expected to be negative.

4. J. N. Spencer, G. M. Bodner, and L. H. Rickard, *Chemistry: Structure & Dynamics*, Third Edition, John Wiley & Sons, 2006. Chapter 13: Problems: 2-4, 6-8.

ChemActivity 53

Entropy (II)

Recall that usually, although not always, exothermic reactions occur naturally (spontaneously). The temperature can also have an impact on whether or not a particular process occurs naturally. This temperature effect is related to the concept of entropy. In fact, it is the entropy change, ΔS, which must be considered in helping to determine whether or not a process occurs naturally.

Model 1: The Melting of Ice.

$$H_2O(s) \rightleftarrows H_2O(l) \quad (T = 273 \text{ K})$$

Critical Thinking Questions

1. Recall that ΔH is positive for the melting of ice. In general, does $\Delta H > 0$ tend to be the case for naturally occurring processes?

2. Do you expect ΔS to be positive or negative for the melting of ice? Explain your reasoning.

3. Recall that ΔH is negative for the freezing of water. In general, does $\Delta H < 0$ tend to be the case for naturally occurring processes?

4. Do you expect ΔS to be positive or negative for the freezing of water? Explain your reasoning.

Model 2: Relationships Between ΔH, ΔS, and T for a Chemical Process to Be Naturally Occurring

ΔH	ΔS	Occurs at Higher T?	Occurs at Lower T?
< 0	< 0	no	yes
< 0	> 0	yes	yes
> 0	< 0	no	no
> 0	> 0	yes	no

Critical Thinking Questions

5. a) Which row in Model 2 corresponds to the melting of ice?

   b) Which row in Model 2 corresponds to the freezing of water?

   c) Confirm that your analysis of the melting of ice and the freezing of water is consistent with the results presented in Model 2.

6. Explain how information presented in Model 2 is consistent with the statement that exothermic reactions tend to occur naturally.

7. a) Based on the information in Model 2, for what values of ΔS do chemical reactions generally occur naturally?

   b) Explain how this is or is not consistent with the previous statement that "chemical (and physical) processes tend to be driven toward the state of highest entropy".

8. Under what conditions can an endothermic reaction occur naturally? Explain your reasoning.

Exercises

1. Why is my desk so messy?

2. When $NH_4NO_3(s)$ dissolves in water the temperature of the solution decreases. Which factor, enthalpy or entropy, makes this a naturally occurring process?

3. J. N. Spencer, G. M. Bodner, and L. H. Rickard, *Chemistry: Structure & Dynamics*, Third Edition, John Wiley & Sons, 2006. Chapter 13: Problems: 9-12.

ChemActivity 54

Entropy Changes in Chemical Reactions

The entropy of atom combination, ΔS°_{ac}, is the change in entropy when a mole of a substance is produced from its constituent atoms in the gas phase at 1 atmosphere pressure and 25°C.

Model 1: The Entropy of Atom Combination, ΔS°_{ac}, of $NO_2(g)$ at 25°C.

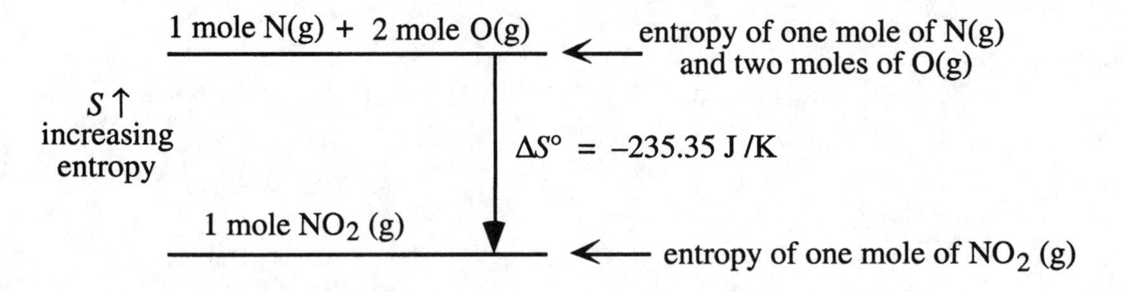

ΔS°_{ac} of $NO_2(g)$ = –235.35 J/mol•K

Table 1. Standard state entropies of atom combination, ΔS°_{ac}.

Substance	ΔS°_{ac} (J/mol•K)	Substance	ΔS°_{ac} (J/mol•K)
N(g)	0	$H_2O(g)$	–202.23
O(g)	0	$H_2O(l)$	–320.57
$N_2(g)$	–114.99	$CCl_4(g)$	–509.04
$O_2(g)$	–116.972	$CCl_4(l)$	–602.49
$NO_2(g)$	–235.35	$C_6H_6(g)$	–1367.7
$N_2O_4(g)$	–646.53	$C_6H_6(l)$	–1464.1

Critical Thinking Questions

1. Do you expect ΔS for the following reaction to be positive or negative? Explain.

$$N(g) + 2\,O(g) \rightleftarrows NO_2(g)$$

2. Why is ΔS°_{ac} of N(g) = 0?

3. Why are the entropies of atom combination of $NO_2(g)$ and $N_2O_4(g)$ negative?

4. For molecules, why are all of the values for entropies of atom combination negative?

5. Why do the entropies of atom combination generally become more negative as the number of atoms in the molecule increases?

6. Why are the entropies of atom combination more negative for liquids than the corresponding entropies of atom combination for gases?

7. Based on the data in Table 1, what is the entropy change associated with breaking one mole of $N_2O_4(g)$ into its constituent atoms (under standard conditions)? Consider both the magnitude and the sign associated with this transformation.

8. Based on the data in Table 1:

   a) What is the entropy change associated with forming one mole of $NO_2(g)$ (under standard conditions) from its constituent atoms?

   b) What is the entropy change associated with forming 2 moles of $NO_2(g)$ from its constituent atoms?

9. Do you expect ΔS for the following reaction to be positive or negative? Explain.

$$N_2O_4(g) \rightleftarrows 2\,NO_2(g)$$

Information

When a chemical reaction takes place, the entropy associated with the chemical system can increase, decrease, or remain constant. This change in entropy can be determined in a manner analogous to the approach that we have taken in calculation of changes in enthalpy, ΔH. That is, we can consider the entropy changes associated with the breaking of the reactants into their constituent gas phase atoms, and the entropy changes associated with reassembling the gas phase atoms to form the products. By combining the entropy changes associated with these processes, the overall ΔS for the reaction can be determined.

To perform this calculation, we need a table of values of the entropies of atom combination for a variety of species. We have seen that the entropy change associated with a process can depend upon temperature and pressure. In addition, entropy changes depend on concentration changes that occur in solution, Thus, we need to define a set of reference conditions, called **standard state conditions**, at which measurements are made. By convention, the standard state conditions for thermodynamic measurements are:

- $T = 298$ K
- All gases have partial pressure of 1 atm.
- All solutes have concentrations of 1 M.

When the change in entropy for a chemical reaction system is measured under these conditions, the result is the **standard state entropy of reaction, $\Delta S°$**.

Model 2: The Entropy Diagram for the Chemical Reaction:

$N_2O_4(g) \rightleftarrows 2NO_2(g)$

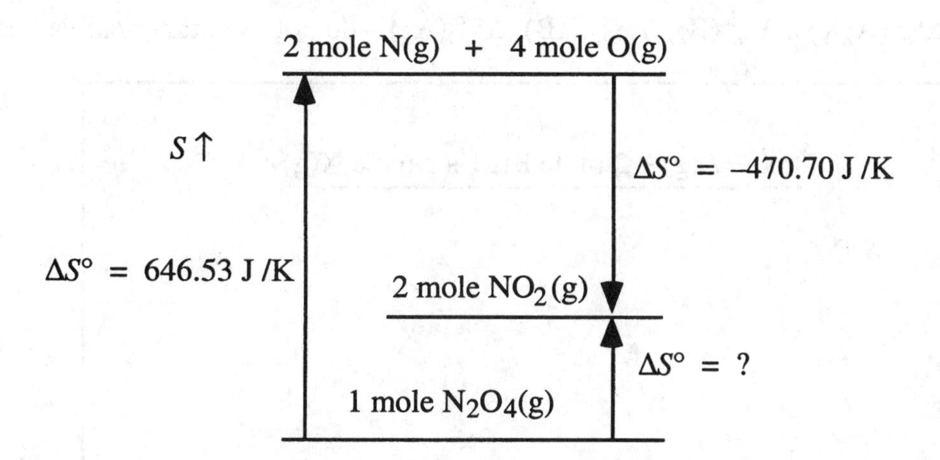

Critical Thinking Questions

10. a) Why is $\Delta S°$ associated with the upward arrow (left-side of Model 2) a positive number?

    b) How was the magnitude of $\Delta S°$ associated with the upward arrow determined?

11. a) Why is $\Delta S°$ associated with the downward arrow (right-side of Model 2) a negative number?

    b) How was the magnitude of $\Delta S°$ associated with the downward arrow determined?

12. Use the data in Model 2 to calculate the $\Delta S°$ for the following reaction:

$N_2O_4(g) \rightleftarrows 2\,NO_2(g)$

13. Complete the diagram below, similar to that in Model 2, to depict $\Delta S°$ for the reaction:

$$A_2X_2(g) + B_2(g) \rightleftarrows 2\,XB(g) + A_2(g)$$

using $\Delta S^{\circ}_{ac}(A_2X_2)$, $\Delta S^{\circ}_{ac}(B_2)$, $\Delta S^{\circ}_{ac}(XB)$, $\Delta S^{\circ}_{ac}(A_2)$—do not use numerical values.

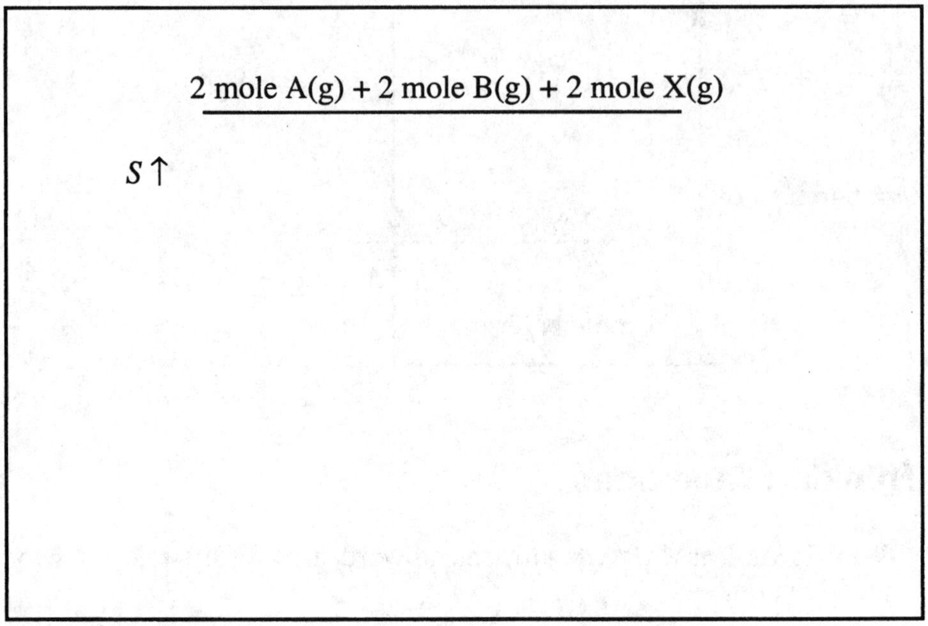

2 mole A(g) + 2 mole B(g) + 2 mole X(g)

$S\uparrow$

14. Using grammatically correct sentences, describe how to calculate the $\Delta S°$ for the reaction in CTQ 13 using the values of ΔS°_{ac} of the four species.

Exercises

1. For each of the following reactions, predict whether $\Delta S°$ will be positive, negative, or zero. Then, calculate $\Delta S°$ for each and compare the value to your predictions.

   a) $N_2(g) + 3\ H_2(g) \rightleftarrows 2\ NH_3(g)$
   b) $4\ Al(s) + 3\ O_2(g) \rightleftarrows 2\ Al_2O_3(s)$
   c) $2\ HCl(g) \rightleftarrows H_2(g) + Cl_2(g)$
   d) $P_4(g) \rightleftarrows 2\ P_2(g)$
   e) $3\ O_2(g) \rightleftarrows 2\ O_3(g)$
   f) $2\ Fe(s) + \frac{3}{2}O_2(g) \rightleftarrows Fe_2O_3(s)$

2. Indicate whether each of the following statements is true or false and explain your reasoning.

   a) The standard state entropy of atom combination for any diatomic gaseous molecule is negative because the formation of a bond is an exothermic process.
   b) The entropy of atom combination for $CH_4(g)$ is expected to be more negative than the entropy of atom combination for $NH_3(g)$.

3. Which ΔS°_{ac} given below is obviously incorrect?

   a) Hg(l) ΔS°_{ac} = +34.71 J/mol•K
   b) P(g) ΔS°_{ac} = 0
   c) $N_2(g)$ ΔS°_{ac} = –122.10 J/mol•K

4. One of the key steps toward transforming coal into a liquid fuel involves the reaction of carbon monoxide with hydrogen to form liquid methanol:

$$CO(g) + 2\ H_2(g) \leftrightarrows CH_3OH(l)$$

   a) Calculate $\Delta S°$ for this reaction.
   b) Provide the oxidation numbers for each of the atoms in these species, and explain whether or not this reaction is an oxidation-reduction process.

5. For the following reaction at 25°C: $Fe_2O_3(s) + 2\ Al(s) \leftrightarrows Al_2O_3(s) + 2\ Fe(s)$ a) Determine the values of $\Delta S°$ and $\Delta H°$. b) Is the reaction favorable or unfavorable with respect to the entropy factor? c) Is the reaction favorable or unfavorable with respect to the enthalpy factor? d) Is this reaction an oxidation-reduction process? Explain.

6. J. N. Spencer, G. M. Bodner, and L. H. Rickard, *Chemistry: Structure & Dynamics*, Third Edition, John Wiley & Sons, 2006. Chapter 13: Problems: 22-25.

Problem

1. Which of the following processes should have the most positive $\Delta S°$? Explain.
   a) $N_2(g) + O_2(g) \leftrightarrows 2NO(g)$
   b) $H_2O(g) \leftrightarrows H_2O(l)$
   c) $3\ C_2H_2(g) \leftrightarrows C_6H_6(l)$
   d) $4\ Al(s) + 3\ O_2(g) \leftrightarrows 2Al_2O_3(s)$
   e) $2\ H_2(g) + O_2(g) \leftrightarrows H_2O(g)$

ChemActivity 55

The Equilibrium Constant (II)

(What Determines the Magnitude of the Equilibrium Constant?)

Model 1: Lower Enthalpy and Higher Entropy are Driving Forces for Chemical Reactions.

We have seen that when the products are at a lower enthalpy than the reactants ($\Delta H° < 0$), a chemical reaction is energetically favored. We have also seen that when the products are more disordered than the reactants ($\Delta S > 0$), a chemical reaction is entropically favored.

Table 1. Standard state enthalpy changes and entropy changes (at 25°C) for several chemical reactions.

Reaction	$\Delta H°$ (kJ/mol_{rxn})	Enthalpy Favorable ?	$\Delta S°$ ($J/mol_{rxn} \cdot K$)	Entropy Favorable ?
$NaCl(s) \rightleftarrows Na^+(aq) + Cl^-(aq)$	3.86		43.3	
$NH_4NO_3(s) \rightleftarrows NH_4^+(aq) + NO_3^-(aq)$	28.07		108.6	
$Zn(s) + Cu^{2+}(aq) \rightleftarrows Cu(s) + Zn^{2+}(aq)$	–218.67		–21.0	
$2Cl^-(1M) + Br_2(l) \rightleftarrows 2Br^-(aq) + Cl_2(g)$	91.23		106.6	
$CH_3COOH(aq) \rightleftarrows CH_3COO^-(aq) + H^+(aq)$	–0.25		–92.0	

Critical Thinking Questions

1. For each of the reactions in the Table 1:

   a) According to the sign of $\Delta H°$, is the reaction favorable or unfavorable with respect to the enthalpy factor? Enter Y or N in the table.

   b) According to the sign of $\Delta S°$, is the reaction favorable or unfavorable with respect to the entropy factor? Enter Y or N in the table.

2. Is there any reaction in Table 1 for which both the enthalpy and the entropy factors are favorable?

3. Is there any reaction in the Table 1 for which both the enthalpy and the entropy factors are unfavorable?

Model 2: Equilibrium Constants and Various Thermodynamic Quantities at 25°C for Several Chemical Reactions.

Reaction	$\Delta H°$ (kJ/mol_{rxn})	$T\Delta S°$ (kJ/mol_{rxn})	$\Delta H° - T\Delta S°$ (kJ/mol_{rxn})	K
$NaCl(s) \rightleftarrows Na^+(aq) + Cl^-(aq)$	3.86	12.9	-9.00	38
$NH_4NO_3(s) \rightleftarrows NH_4^+(aq) + NO_3^-(aq)$	28.07	32.38	-4.31	5.7
$Zn(s) + Cu^{2+}(aq) \rightleftarrows Cu(s) + Zn^{2+}(aq)$	-218.67	-6.3	-212.4	1.6×10^{37}
$2Cl^-(aq) + Br_2(l) \rightleftarrows 2Br^-(aq) + Cl_2(g)$	91.23	31.78	59.45	3.9×10^{-11}
$CH_3COOH(aq) \rightleftarrows CH_3COO^-(aq) + H^+(aq)$	-0.25	-27.43	27.18	1.7×10^{-5}

Critical Thinking Questions

4. According to the value of the equilibrium constants, which reactions in Model 2 appear to be favorable ($K>1$)? Unfavorable ($K<1$)?

5. Which factor in Model 2 is an indicator that $K>1$ or $K<1$: $\Delta H°$, $T\Delta S°$, $\Delta H° - T\Delta S°$?

6. When $K>1$, is the factor identified in CTQ 5 positive or negative?

7. When $K<1$, is the factor identified in CTQ 5 positive or negative?

8. What is the qualitative relationship between the value of the factor identified in CTQ 5 and the magnitude of K?

Exercises

1. For each of the following reactions, use the appropriate tables to determine ΔH° and ΔS° (at 25°C). Then, indicate whether the equilibrium constant is expected to be greater than, less than, or equal to 1, or that it cannot be deduced.

   a) $\frac{1}{8} S_8(s) + O_2(g) \rightleftarrows SO_2(g)$
   b) $2\ C(graphite) + 2\ H_2(g) \rightleftarrows C_2H_4(g)$
   c) $CuO(s) + H_2(g) \rightleftarrows Cu(s) + H_2O(l)$
   d) $N_2(g) + 3\ H_2(g) \rightleftarrows 2\ NH_3(g)$
   e) $H_2O(l) \rightleftarrows H_2O(g)$

2. Use enthalpies of atom combination and entropies of atom combination to determine if any of the following reactions have an equilibrium constant greater than 1. Find the reaction with the greatest equilibrium constant. Find the reaction with the smallest equilibrium constant.

   a) $HF(aq) \rightleftarrows H^+(aq) + F^-(aq)$
   b) $N_2(g) + 3\ H_2(g) \rightleftarrows 2\ NH_3(g)$
   c) $PbCl_2(s) \rightleftarrows Pb^{2+}(aq) + 2\ Cl^-(aq)$

3. Without referring to tables to calculate ΔH° and ΔS°, predict whether the equilibrium constant at room temperature for the following exothermic reaction will be greater than, less than, or equal to 1. Explain your reasoning.

   $C_3H_8(g) + 5\ O_2(g) \rightleftarrows 3\ CO_2(g) + 4\ H_2O(g)$

4. Calculate ΔH° and ΔS° for the reaction:

   $3\ Fe(s) + 4\ H_2O(l) \rightleftarrows Fe_3O_4(s) + 4\ H_2(g)$

   Recalling that hydrogen gas is quite flammable, explain why it is a mistake to use water to put out a fire that contains white-hot iron metal.[1]

5. J. N. Spencer, G. M. Bodner, and L. H. Rickard, *Chemistry: Structure & Dynamics*, Third Edition, John Wiley & Sons, 2006. Chapter 13: Problems: 29, 30, 32, 34, 40-42.

---

[1] J. N. Spencer, G. M. Bodner, and L. H. Rickard, *Chemistry: Structure & Dynamics*, Third Edition, John Wiley & Sons, 2006, Chapter 13, Problem 39.

Model 3: The Gibbs Free Energy.

An equation that describes the quantitative relationship between the enthalpy, the entropy and the equilibrium constant was developed by J. Willard Gibbs, a professor of mathematical physics at Yale in the late nineteeth century. He defined a new quantity, now called the **Gibbs free energy** (G), which describes the balance between the enthalpy and entropy factors for a chemical reaction.

$$G = H - TS.$$

For a chemical reaction which takes place at a constant temperature,

$$\Delta G = G(\text{products}) - G(\text{reactants}) = \Delta H - T\Delta S$$

If the reactants and products are in standard states at 25°C:

$$\Delta G^\circ = G^\circ(\text{products}) - G^\circ(\text{reactants}) = \Delta H^\circ - T\Delta S^\circ$$

Critical Thinking Questions

9. For a chemical reaction with $K > 1$, is ΔG° positive or negative?

10. For a chemical reaction with $K < 1$, is ΔG° positive or negative?

11. If $\Delta H^\circ = T\Delta S^\circ$, what is the value of ΔG°? Predict the value of K in this case.

Exercises

6. For what combination of values of ΔH° and ΔS° will a chemical reaction always have $K < 1$? Always have $K > 1$?

7. J. N. Spencer, G. M. Bodner, and L. H. Rickard, *Chemistry: Structure & Dynamics*, Third Edition, John Wiley & Sons, 2006. Chapter 13: Problems: 44, 95.

ChemActivity 56

The Equilibrium Constant (III)

(How Are $\Delta G°$ and K Related?)

Model 1: The Mathematical Relationship Between $\Delta G°$ and K.

Table 1. Standard state free energy changes and equilibrium constants for several chemical reactions (25°C).

Reaction	$\Delta G°$ (kJ/mol$_{rxn}$)	K
$NaCl(s) \rightleftarrows Na^+(aq) + Cl^-(aq)$	–9.00	38
$NH_4NO_3(s) \rightleftarrows NH_4^+(aq) + NO_3^-(aq)$	–4.31	5.7
$Zn(s) + Cu^{2+}(aq) \rightleftarrows Cu(s) + Zn^{2+}(aq)$	–212.4	1.6×10^{37}
$2Cl^-(aq) + Br_2(l) \rightleftarrows 2Br^-(aq) + Cl_2(g)$	59.45	3.9×10^{-11}
$CH_3COOH(aq) \rightleftarrows CH_3COO^-(aq) + H^+(aq)$	27.18	1.8×10^{-5}

Critical Thinking Questions

1. Which expression below describes the mathematical relationship between $\Delta G°$ and K?

    a) Is $\Delta G° \propto K$? (That is, does $\Delta G° = c\,K$, where "c" is some proportionality constant? If so, then $\frac{\Delta G°}{K}$ = same number for all entries in Table 1.)

    b) Is $\Delta G° \propto -K$?

    c) Is $\Delta G° \propto -\sqrt{K}$?

    d) Is $\Delta G° \propto -\ln K$?

2. What is the value (with units) of the proportionality constant in CTQ 1?

Information

The equation $\Delta G° = -RT\ln K$ is one of the most important equations in chemistry. It relates the change in standard state free energies for a chemical reaction to the equilibrium constant. Thus, it is possible to calculate the value of an equilibrium constant for a reaction *before* the reaction takes place.

Critical Thinking Questions

3. Show that the value of the proportionality constant found in CTQ 2 is equal to RT, where the universal gas constant $R = 8.3145 \frac{\text{J}}{\text{K mole}}$ and T = 25°C.

4. Recall that $\Delta G°$ can be written as a function of $\Delta H°$ and $\Delta S°$. Assume that $\Delta H°$ and $\Delta S°$ are not temperature dependent and answer each of the following:

   a) Derive an expression relating $\ln K$, $\Delta H°$, and $\Delta S°$. That is, derive an expression that looks like $\ln K$ = some function of $\Delta H°$ and $\Delta S°$. The temperature, T, should appear only once in this equation.

   b) How are the equilibrium constants for reactions with $\Delta H° > 0$ affected by an increase in temperature?

   c) How are equilibrium constants for reactions with $\Delta H° < 0$ affected by an increase in temperature?

5. How does temperature affect the equilibrium $H_2O(l) \rightleftarrows H_2O(g)$? Explain in terms of $\Delta H°$.

Model 2: The Mathematical Relationship Between $E°$ and K.

Table 2. Measured voltages and equilibrium constants for some galvanic cells using standard electrodes at 25°C (all ions and soluble species at 1 M and all gases at 1 atm).

Cathode	Anode	$E°$ (V)	K
Cu/Cu^{2+}	Zn/Zn^{2+}	1.10	1.6×10^{37}
Cu/Cu^{2+}	SHE	0.34	3.1×10^{11}
Br_2/Br^-	Zn/Zn^{2+}	1.85	3.5×10^{62}
Zn/Zn^{2+}	K/K^+	2.16	1.1×10^{73}
Cl_2/Cl^-	Ag/Ag^+	0.56	8.6×10^{18}

Critical Thinking Questions

6. Write chemical equations for each of the galvanic cells in Table 2.

7. Which expression below describes the mathematical relationship between $E°$ and K?

a) Is $E° \propto K$?

b) Is $E° \propto -K$?

c) Is $E° \propto \sqrt{K}$?

d) Is $E° \propto \ln K$?

8. What is the value (with units) of the proportionality constant in CTQ 7?

Information

The equation $E° = \frac{RT}{nF}\ln K$ relates the standard state cell potential for a chemical reaction to the equilibrium constant. Thus, it is possible to determine the value of an equilibrium constant for a reaction by measurement of the cell potential.

Critical Thinking Questions

9. Show that the proportionality constant found in CTQ 7 is equal to $\frac{RT}{nF}$, where the universal gas constant $R = 8.314\frac{\text{J}}{\text{K mole}}$, F is Faraday's constant, 96485 coulombs per mole (of electrons), and "n" is the number of moles of electrons transferred in the balanced chemical reaction.

10. Using the equations in the Information sections of Model 1 and Model 2, derive an equation that relates $\Delta G°$ and $E°$.

Exercises

1. Calculate $\Delta H°$, $\Delta S°$, and $\Delta G°$ for the reaction of $NH_3(aq)$ with $H_2O(l)$ to produce $NH_4^+(aq)$ and $OH^-(aq)$. Use these data to calculate K_b for ammonia.

2. Consider the reaction $CO_2(g) + H_2(g) \rightleftarrows CO(g) + H_2O(g)$. Calculate $\Delta H°$, $\Delta S°$, $\Delta G°$, and K for this reaction at 25°C. Predict the effect on the equilibrium constant when the temperature of the system is increased.

3. Without referring to the tables of thermodynamic data, predict the signs of $\Delta H°$ and $\Delta S°$ for the reaction $NH_3(aq) \rightleftarrows NH_3(g)$. Explain why the odor of $NH_3(g)$ that collects above an aqueous solution of ammonia becomes more intense as the temperature is increased.[1]

4. Explain why the equilibrium constant for the reaction:

$$N_2(g) + 3\,H_2(g) \rightarrow 2\,NH_3(g)$$

decreases as the temperature increases.

---

[1] J. N. Spencer, G. M. Bodner, and L. H. Rickard, *Chemistry: Structure & Dynamics*, Third Edition, John Wiley & Sons, 2006, Chapter 13, Problem 31.

5. Assume that a liquid boils at the temperature at which $\Delta G° = 0$ for the reaction liquid $\rightarrow$ gas.

   a) Adding salt to water does not change $\Delta H°$ for the process

      $H_2O(\ell) \rightarrow H_2O(g)$

      However, $\Delta S°$ for the process is decreased because the entropy of the liquid is increased without changing the entropy of the gas. Show how this can be used to explain the fact that adding salt to water raises its boiling point.

   b) Using the appropriate values of $\Delta H°$and $\Delta S°$, estimate the boiling point of methanol (CH_3OH).

6. Consider the reaction:

   $$PbCl_2(s) \rightleftarrows Pb^{2+}(aq) + 2\ Cl^{-}(aq) \quad K = 1.6 \times 10^{-5}$$

   a) Calculate $\Delta G°$ for this reaction at 298 K.

   b) Predict the signs of $\Delta S°$ and $\Delta H°$ for this reaction. Explain your reasoning.

7. Calculate the equilibrium constant at 25°C for each of the following reactions (from the standard cell potential).

   a) $Zn(s) + 2\ H^{+}(aq) \rightleftarrows H_2(g) + Zn^{2+}(aq)$
   b) $2\ Na(s) + 2\ H_2O(l) \rightleftarrows H_2(g) + 2\ Na^{+}(aq) + 2\ OH^{-}(aq)$
   c) $Cu(s) + 2\ H^{+}(aq) \rightleftarrows H_2(g) + Cu^{2+}(aq)$

8. J. N. Spencer, G. M. Bodner, and L. H. Rickard, *Chemistry: Structure & Dynamics*, Third Edition, John Wiley & Sons, 2006. Chapter 13: Problems: 45, 72-76, 78-82, 88, 90, 92.

Problem

1. A voltaic cell has the following overall reaction:

   $$I_3^{-}(aq) + 2\ S_2O_3^{2-}(aq) \rightleftarrows 3\ I^{-}(aq) + S_4O_6^{2-}(aq)$$

   a) Determine the cell voltage, $E°$, when run under standard conditions.
   b) Which chemical species is the oxidizing agent?
   c) Determine the value of the equilibrium constant for this reaction at 25°C.

ChemActivity 57

Rates of Chemical Reactions (II)

(How Does the Concentration of Reactants Affect the Rate?)

Model 1: The Rate of a Reaction Varies with Time.

We have previously defined the rate of reaction as

$$\text{rate} = -\frac{\Delta(\text{reactant})}{\Delta\text{time}}$$

for any chemical reactant with a stoichiometric coefficient of 1 in the balanced chemical equation.

A better measure of the rate of a reaction is the *instantaneous rate of reaction,* generally written as

$$\text{rate} = -\frac{d(\text{reactant})}{dt}$$

The value of the instantaneous rate of reaction (for reactants with a stoichiometric coefficient of one in the balanced chemical equation) can be obtained by plotting the concentration of the reactant versus time, drawing a tangent to the curve, and determining the slope of the line.

Figure 1. Nitrite concentration versus time for the reaction of ammonium ion with nitrite ion.

$$NH_4^+(aq) + NO_2^-(aq) \rightleftarrows N_2(g) + 2\,H_2O(l)$$

$$(NO_2^-)_o = 0.00500 \text{ M} \qquad (NH_4^+)_o = 0.100 \text{ M}$$

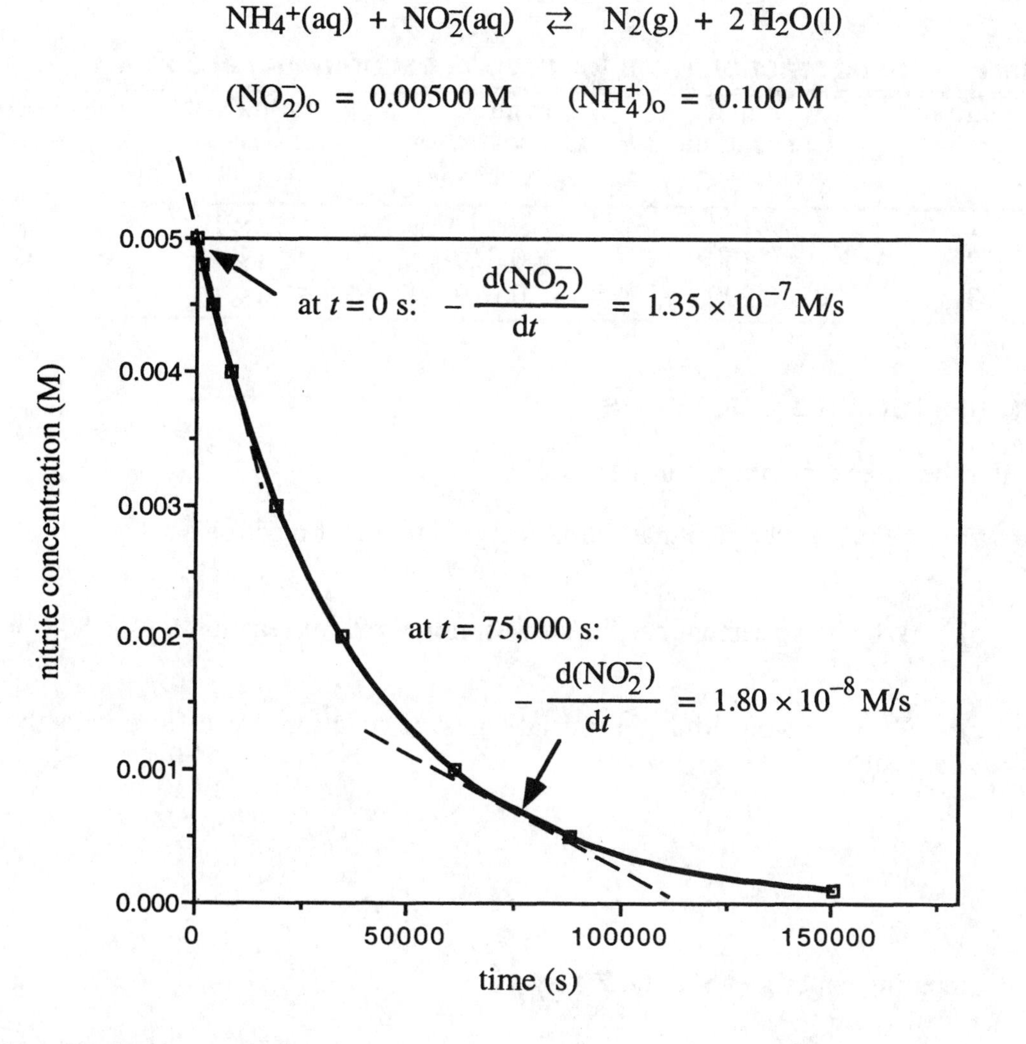

Critical Thinking Questions

1. What is the rate of reaction at $t = 0$ s?

2. What is the rate of reaction at $t = 75{,}000$ s?

3. How does the rate of reaction change as (NO_2^-) decreases?

4. Estimate the value of the rate of reaction at $t = 175{,}000$ s. Explain your reasoning.

Model 2: The Effect of Concentration on Reaction Rate.

$$NH_4^+(aq) + NO_2^-(aq) \rightleftarrows N_2(g) + 2\,H_2O(l)$$

Table 1. Initial reaction rates for several experiments at 25°C.

Experiment	Initial Concentration of NH_4^+ (M)	Initial Concentration of NO_2^- (M)	Initial Rate of Reaction (M /sec)
1	0.100	0.0050	1.35×10^{-7}
2	0.100	0.010	2.70×10^{-7}
3	0.200	0.010	5.40×10^{-7}

Critical Thinking Questions

5. For the three experiments in Table 1:

   a) Which experiment has the fastest initial rate of reaction?

   b) Which experiment has the slowest initial rate of reaction?

   c) Why do you think the initial rates of reaction are different in the three experiments?

6. Comparing experiments **1** and **2** only:

   a) Are the initial concentrations of NH_4^+ the same? If not, what is the ratio of the concentrations expressed as a fraction, $(NH_4^+)_2/(NH_4^+)_1$?

   b) Are the initial concentrations of NO_2^- the same? If not, what is the ratio of the concentrations expressed as a fraction, $(NO_2^-)_2/(NO_2^-)_1$?

   c) Are the initial rates of reaction the same? If not, what is the ratio of the rates of reaction expressed as a fraction, initial $rate_2$/initial $rate_1$?

   d) Based on the answers to parts a)–c) above, can you determine whether or not the initial rate of reaction depends on the initial (NH_4^+)? Why or why not?

e) Based on the answers to parts a)–c) above, can you determine whether or not the initial rate of reaction depends on the initial (NO_2^-)? Why or why not?

f) Based on your answers to a)–c), does the rate of reaction appear to be proportional to (NO_2^-) raised to some power? If so, what is the power?

7. Comparing experiments **2** and **3** only:

a) Are the initial concentrations of NH_4^+ the same? If not, what is the ratio of the concentrations expressed as a fraction, $(NH_4^+)_3/(NH_4^+)_2$?

b) Are the initial concentrations of NO_2^- the same? If not, what is the ratio of the concentrations expressed as a fraction, $(NO_2^-)_3/(NO_2^-)_2$?

c) Are the initial rates of reaction the same? If not, what is the ratio of the rates of reaction expressed as a fraction, initial $rate_3$/initial $rate_2$?

d) Based on the answers to parts a)–c) above, can you determine whether or not the initial rate of reaction depends on the initial (NH_4^+)? Why or why not?

e) Based on the answers to parts a)–c) above, can you determine whether or not the initial rate of reaction depends on the initial (NO_2^-)? Why or why not?

f) Based on your answers to a)–c), does the rate of reaction appear to be proportional to (NH_4^+) raised to some power? If so, what is the power?

Model 3: The Rate Law.

Often the rate of reaction is found to be proportional to the concentration of a reactant raised to some power (usually an integer such as 0, 1, 2, ...). For example, if

$$\text{rate} = k\,(R)^x \qquad \text{then,}$$

$$\frac{\text{initial rate}_2}{\text{initial rate}_1} = \frac{k\,(R)_2^x}{k\,(R)_1^x} = \left(\frac{(R)_2}{(R)_1}\right)^x$$

where initial rate_i = the initial rate of experiment i
$(R)_i$ = the initial concentration of the reactant R for experiment i

The relationship between the rate of a reaction and the concentrations of reactants is known as the **rate law.** An example of a typical rate law is

$$\text{rate} = k\,(NH_4^+)^x\,(NO_2^-)^y \tag{1}$$

where k is the proportionality constant, known as the **rate constant**, and x and y are the exponents described previously. The rate constant is characteristic of a particular reaction at a given temperature. The exponents are often referred to as the order of the reaction with respect to the respective reactants. For example, if $x = 3$, we say that the reaction is third order with respect to NH_4^+. The rate constant and the exponents (or orders) can be determined by experiment only.

Critical Thinking Questions

8. Based on your answers to CTQ 7, determine the order of the reaction in the Model 2 with respect to NH_4^+.

9. Based on your answers to CTQ 6 determine the order of the reaction in the Model 2 with respect to NO_2^-.

10. a) Based on your answers to CTQs 8 and 9, calculate the value of the rate constant k in the rate law for the reaction using

    i) data from Experiment 1

    ii) data from Experiment 2

    iii) data from Experiment 3

b) Compare the three answers from part a). Explain why the relative values are reasonable.

Table 2. Experimental rate laws for several chemical reactions.

Reaction	Experimental Rate Law
$CH_3Br(aq) + OH^-(aq) \rightleftarrows CH_3OH(aq) + Br^-(aq)$	rate = k (CH_3Br)
$2NO(g) + O_2(g) \rightleftarrows 2\,NO_2(g)$	rate = k $(NO)^2\,(O_2)$
$2\,HI(g) \rightleftarrows H_2(g) + I_2(g)$	rate = k $(HI)^2$
$NH_4^+(aq) + NO_2^-(aq) \rightleftarrows N_2(g) + 2\,H_2O(l)$	rate = k $(NH_4^+)\,(NO_2^-)$
$BrO_3^-(aq) + 5Br^-(aq) + 6H^+(aq) \rightleftarrows 3Br_2(aq) + 3H_2O$	rate = k $(BrO_3^-)(Br^-)\,(H^+)^2$
$CH_3CHO(g) \rightleftarrows CH_4(g) + CO(g)$	rate = k $(CH_3CHO)^{3/2}$

Critical Thinking Questions

11. Based on the data in Table 2, is the order of a reaction with respect to a particular species always equal to its stoichiometric coefficient in the balanced chemical equation?

12. Comment on the appropriateness of the following methods to determine the order of a reaction.

    a) Examine the stoichiometric coefficients in the balanced chemical equation. In this method, the order of a reaction with respect to a component is equal to the stoichiometric coefficient of that component in the balanced chemical equation.

    b) Perform experiments. In this method, the order of a reaction with respect to a component is determined by how the reaction rate changes when the concentration(s) is changed.

Exercises

1. What is the initial rate of production of N_2 in experiment 3 of Table 1?

2. The following initial reaction rates were observed for the oxidation of Fe^{2+} by Ce^{4+}:

Experiment	Initial Concentration of Ce^{4+} (M)	Initial Concentration of Fe^{2+} (M)	Initial Rate of Reaction (M /sec)
1	1.5×10^{-5}	2.5×10^{-5}	3.79×10^{-7}
2	1.5×10^{-5}	5.0×10^{-5}	7.58×10^{-7}
3	3.0×10^{-5}	5.0×10^{-5}	1.52×10^{-6}

a) Determine the order of the reaction with respect to Ce^{4+} and with respect to Fe^{2+}.
b) Write the rate law for this reaction.
c) Calculate the rate constant, k, and give its units.
d) Predict the initial reaction rate for a solution in which (Ce^{4+}) is 1.0×10^{-5} M and (Fe^{2+}) is 1.8×10^{-5} M.

3. Determine the rate law and evaluate the rate constant for the following reaction:

$$2\,NO(g) + Br_2(g) \rightleftarrows 2\,NOBr\,(g)$$

EXP	$(NO)_o$ (M)	$(Br_2)_o$ (M)	Initial Rate of Reaction (M/min)
1	0.10	0.10	1.30×10^{-3}
2	0.20	0.10	5.20×10^{-3}
3	0.20	0.30	1.56×10^{-2}

4. The following reaction was studied experimentally at 25°C.

$$S_2O_8^{2-}(aq) + 2I^-(aq) \rightleftarrows I_2(aq) + 2SO_4^{2-}(aq)$$

The reaction was found to be first order in I^- and first order in $S_2O_8^{2-}$. A reaction was run with $(I^-)_o = 0.080$ M and $(S_2O_8^{2-})_o = 0.040$ M. The initial rate of formation of I_2 was found to be $1.25 \times 10^{-6} \frac{\text{mole}}{\text{liter s}}$. Provide an expression for the rate law for this reaction, and determine the initial rate of formation of I_2 when $(I^-)_o = 0.080$ M and $(S_2O_8^{2-})_o = 0.060$ M.

5. Indicate whether the following statement is true or false and explain your reasoning.

   The rate law for a reaction can be obtained by examining the balanced chemical equation for the reaction.

6. J. N. Spencer, G. M. Bodner, and L. H. Rickard, *Chemistry: Structure & Dynamics*, Third Edition, John Wiley & Sons, 2006. Chapter 14: Problems: 45-50.

Problems

1. One of the major irritants found in smog is formaldehyde, $CH_2O(g)$, formed by the reaction of ethene and ozone in the atmosphere:

$$C_2H_4(g) + 2\,O_3(g) \rightleftarrows 4\,CH_2O(g) + O_2(g)$$

From the following initial rate data, deduce the rate law for this reaction. Clearly indicate how you arrived at your answer.

Experiment	Initial Concentration of O_3 (M)	Initial Concentration of C_2H_4 (M)	Initial Rate of Reaction (M /sec)
1	0.5×10^{-7}	1.0×10^{-8}	1.0×10^{-12}
2	1.5×10^{-7}	1.0×10^{-8}	3.0×10^{-12}
3	1.0×10^{-7}	2.0×10^{-8}	4.0×10^{-12}

2. For the following reaction:

$$2\,HgCl_2(aq) + C_2O_4^{2-}(aq) \rightleftarrows Hg_2Cl_2(s) + 2\,Cl^-(aq) + 2\,CO_2(g)$$

Experiment	Initial Concentration of $HgCl_2$ (M)	Initial Concentration of $C_2O_4^{2-}$ (M)	Initial Rate of Reaction (M /sec)
1	0.096	0.13	2.1×10^{-7}
2	0.096	0.21	5.5×10^{-7}
3	0.171	0.21	9.8×10^{-7}

   a) Determine the order of the reaction with respect to $HgCl_2$ and with respect to $C_2O_4^{2-}$.
   b) Write the rate law for this reaction.
   c) Calculate the rate constant and give its units.

3. Consider the reaction

$$2\,UO_2^+(aq) + 4\,H^+(aq) \rightleftarrows U^{4+}(aq) + UO_2^{2+}(aq) + 2H_2O(l)$$

   a) From the following initial rate data, deduce the rate law for this reaction. Clearly indicate how you arrived at your answer. b) Find the rate constant *k*, including units, for the reaction above.

Experiment	Initial Concentration of UO_2^+ (M)	Initial Concentration of H^+ (M)	Initial Rate of Reaction (M /sec)
1	0.0012	0.22	4.12×10^{-5}
2	0.0012	0.35	6.55×10^{-5}
3	0.0030	0.35	4.10×10^{-4}

4. The following data were collected for the reaction:

$$2\,NO(g) + O_2(g) \rightleftarrows 2\,NO_2(g)$$

Initial NO Concentration (mol/L)	Initial O_2 Concentration (mol/L)	Initial Rate of reaction (mol/Ls)
5.38×10^{-3}	5.38×10^{-3}	1.91×10^{-5}
8.07×10^{-3}	5.38×10^{-3}	4.30×10^{-5}
13.45×10^{-3}	5.38×10^{-3}	11.94×10^{-5}
8.07×10^{-3}	6.99×10^{-3}	5.59×10^{-5}
8.07×10^{-3}	9.69×10^{-3}	7.75×10^{-5}

What is the rate law for the reaction?

ChemActivity 58

Integrated Rate Laws

(How Does the Concentration of a Reactant Change as the Reaction Proceeds?)

Model 1: Integrated First- and Second-Order Rate Laws

As we have seen, the concentration of a reactant decreases as a reaction proceeds. In a few situations, the concentration of a reactant is a (relatively) simple function of time. For example, for a reaction that is first order in a single reactant, R, the rate law is

$$\text{rate} = -\frac{d(R)}{dt} = k\,(R)^1$$

This equation can be rearranged and integrated to provide an explicit relationship between (R) and time. The integrated form of a first-order rate law is

$$\ln (R) = \ln (R)_o - kt$$

where $(R)_o$ is the (R) at time 0 (the initial concentration) and t is the time.

Similarly, for a second-order reaction with the rate law

$$\text{rate} = -\frac{d(R)}{dt} = k\,(R)^2$$

the corresponding relationship is

$$\frac{1}{(R)} = \frac{1}{(R)_o} + kt$$

Note that the integrated rate laws contain four potential variables: (R), $(R)_o$, k, and t. Knowledge of any three of these variables permits the calculation of the fourth variable.

Critical Thinking Questions

1. For both integrated rate laws, show that (R) decreases as t increases.

2. If one obtained data for a first-order reaction, and then made a graph of ln(R) (along the vertical axis) versus t (along the horizontal axis), the resulting plot would be a straight line.

    a) What is the slope?

    b) What is the y-intercept?

(Hint: Compare the first-order integrated rate law to the equation $y = mx + b$.)

3. If one obtained data for a second-order reaction, and then made a graph of 1/(R) versus *t*, the resulting plot would be a straight line.

   a) What would be the slope?

   b) What would be the *y*-intercept?

Exercises

1. Consider the decomposition of N_2O_5 in the gas phase:

   $$N_2O_5(g) \rightarrow 2\,NO_2(g) + \frac{1}{2}\,O_2(g)$$

   At room temperature, the following data were collected.

   Concentration of N_2O_5 as a function of time.

Time (s)	(N_2O_5) (M)
0	0.1000
50	0.0707
100	0.0500
200	0.0250
400	0.00625

   a) Copy these data into an appropriate computer file, and construct two plots to determine whether it is consistent with a first-order rate law or a second-order rate law. Once you have determined which rate law applies, determine the value of the rate constant.

   b) What is the concentration of $N_2O_5(g)$ at $t = 10$ s?

2. The chemical equation for the decomposition of hypobromite ion, BrO^-, is:

   $$3\,BrO^-(aq) \rightleftarrows BrO_3^-(aq) + 2\,Br^-(aq)$$

   The concentration of hypobromite was monitored as a function of time, shown below:

time (s)	BrO^- conc (M)
0	0.750
20	0.408
40	0.280
60	0.213
80	0.172
100	0.144

   Explain, in detail, how you would proceed to determine if the reaction was first-order or second-order.

3. The isomerization reaction

   $CH_3NC(g) \rightleftarrows CH_3CN(g)$

   obeys the first-order rate law. At 500 K, the concentration of CH_3NC is 85% of its original value after 247 s. What is the rate constant for this decomposition at 500 K? At what time will the concentration of CH_3NC be 25% of its original value?

4. Indicate whether the following statement is true or false and explain your reasoning:

   For the first-order reaction A → products, the rate of reaction remains constant as the reaction proceeds.

5. Which graph (I, II, III, IV, V) best describes the following reaction if the reaction is first order in N_2O_4?[1]

   $N_2O_4(g) \rightarrow 2\ NO_2(g)$

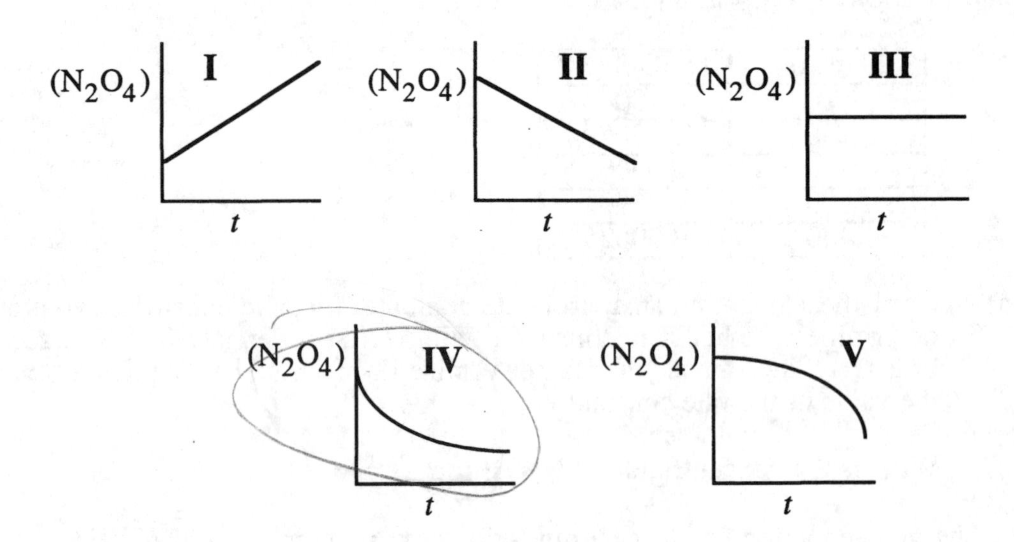

6. J. N. Spencer, G. M. Bodner, and L. H. Rickard, *Chemistry: Structure & Dynamics*, Third Edition, John Wiley & Sons, 2006. Chapter 14: Problems: 51, 54, 55, 101.

[1] J. N. Spencer, G. M. Bodner, and L. H. Rickard, *Chemistry: Structure & Dynamics*, Third Edition, John Wiley & Sons, 2006, Chapter 14, Problem 56.

Model 2: A Simple Decomposition Reaction.

Chloroethane decomposes at 800K:

$$CH_3CH_2Cl(g) \rightleftarrows C_2H_4(g) + HCl(g)$$

The reaction is first-order with respect to chloroethane.

Figure 1. The concentration of chloroethane versus time at 800 K.

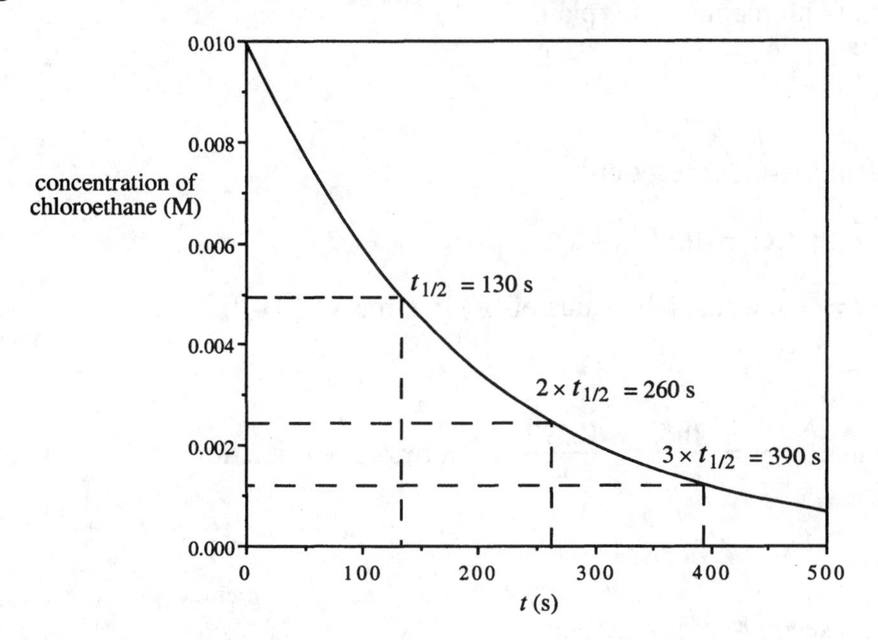

Critical Thinking Questions

4. In Figure 1:

   a) What is the initial concentration of chloroethane?

   b) What is the concentration of chloroethane at $t = 130$ s? What is the ratio of the concentration of chloroethane at $t = 130$ s to the concentration of chloroethane at $t = 0$ s?

   c) What is the concentration of chloroethane at $t = 260$ s? What is the ratio of the concentration of chloroethane at $t = 260$ s to the concentration of chloroethane at $t = 130$ s?

   d) What is the concentration of chloroethane at $t = 390$ s? What is the ratio of the concentration of chloroethane at $t = 390$ s to the concentration of chloroethane at $t = 260$ s?

5. Use Figure 1 to estimate the concentration of chloroethane at $t = 50$ s. Estimate the concentration of chloroethane at $t = 180$ s. Is this result consistent with your answers to CTQ 4? Explain.

Model 3: The Half-Life of a Reaction.

The half-life of a reaction, $t_{1/2}$, is the time that it takes for the concentration of a single reactant to reach one-half of its original value.

Critical Thinking Questions

6. Based on your answers to CTQs 4 and 5, does $t_{1/2}$ for the reaction in Model 2 on the concentration of chloroethane? Explain.

7. Recall that for a first-order reaction:

$$\ln (R) = \ln (R)_o - kt$$

a) When $t = t_{1/2}$, what is the value of (R) in terms of $(R)_o$?

$R = \frac{1}{2} R_o = \frac{R}{2}$

b) Show that $t_{1/2} = \frac{\ln 2}{k} = \frac{0.693}{k}$ for a first-order reaction.

$\ln (R) = \ln_o - kt$

$\ln(R) - \ln(R)_o = kt$

8. Recall that for a second-order reaction:

$$\frac{1}{(R)} = \frac{1}{(R)_o} + kt$$

a) When $t = t_{1/2}$, what is the value of (R) in terms of $(R)_o$?

$R = \frac{(R)_o}{2}$

b) Show that $t_{1/2} = \frac{1}{k\,(R)_o}$ for a second-order reaction.

$\frac{1}{\frac{R}{2}} = \frac{1}{R_o} + kt$

$\frac{1}{R_o k}$

Exercises

7. The isomerization reaction

$$CH_3NC(g) \rightleftarrows CH_3CN(g)$$

obeys a first-order rate law at 500 K. See Ex. 3 for the rate constant at 500 K. What is the half-life of this decomposition? How long will it take for the concentration of CH_3NC to reach 25% of its original value? How well does this answer agree with your answer in Ex. 3?

8. What fraction of reactant remains after 3 half-lives of a first-order reaction?
a) 1/2 b) 1/3 c) 1/6 d) 1/8 e) 1/12

9. One way to determine the age of a rock is to measure the extent to which the ^{87}Rb in the rock has decayed to ^{87}Sr (a first-order process).

$$^{87}Rb \rightleftarrows {}^{87}Sr + e^- \qquad k = 1.42 \times 10^{-11} \text{ year}^{-1}$$

What fraction of the original ^{87}Rb would still remain in the rock after 1.0×10^{10} years (10 billion years)?

10. The radioactive decay of ^{14}C is a first-order process with a half-life of 5730 years. If living wood gives 15.3 disintegrations per minute per gram, and a wooden bowl found in an archeological dig gives 6.29 disintegrations per minute per gram, how old is the bowl?

11. J. N. Spencer, G. M. Bodner, and L. H. Rickard, *Chemistry: Structure & Dynamics*, Third Edition, John Wiley & Sons, 2006. Chapter 14: Problems: 57, 58, 60-62, 64, 67.

Problems

1. The exothermic reaction $2\ A_2B_2(g) \rightleftarrows 2\ A_2(g) + 2\ B_2(g)$ has the experimental rate law: rate = $k\ (A_2B_2)^2$. Explain how the rate constant for this reaction can be determined from experimental measurements of (A_2B_2) at 10 minute intervals. (That is, from experimental determinations of (A_2B_2) when t = 0 min, t = 10 min, t = 20 min, etc.)

2. The rate law for a reaction is known to involve only the reactant A, and is suspected to be either first-order or second-order. Describe, using grammatically correct English sentences, how the order of the reaction can be determined by measuring how long it takes for the concentration of A to reach 50% and 25% of its original value.

3. The reaction A $\rightleftarrows$ B + C is known to follow a first-order rate law. What feature of the plot of the concentration of A vs. time (shown below) clearly indicates that the reaction is indeed first-order, not second-order.

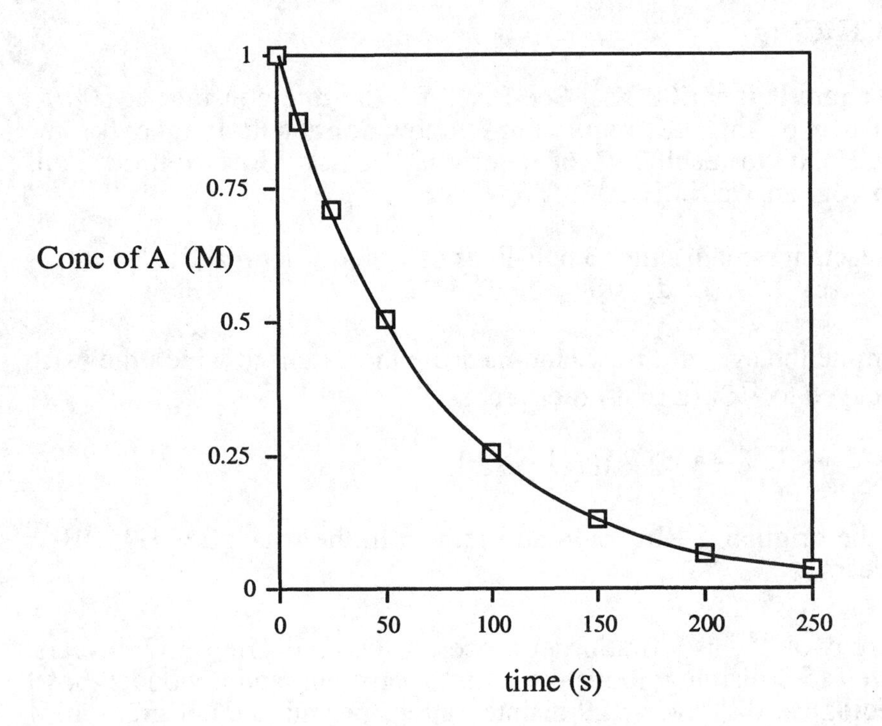

ChemActivity 59

Reaction Mechanisms (I)

(How Fast Will a Reaction Be?)

Model 1: Eight Balls in a Lopsided, Double-box.

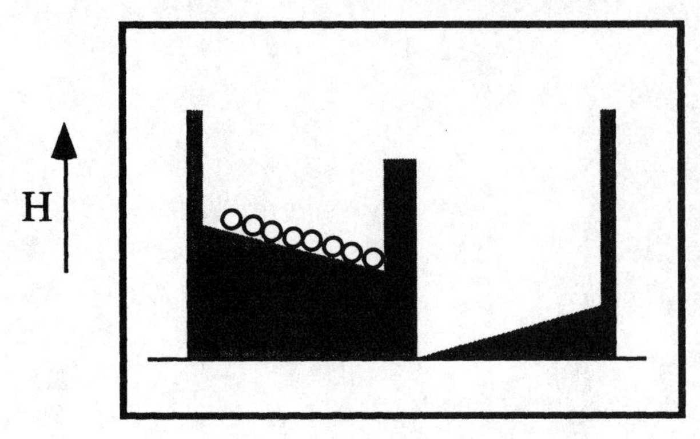

state (i)

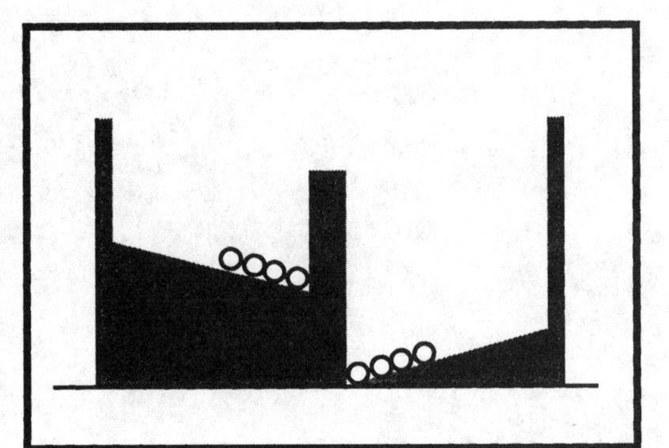

state (ii)

Critical Thinking Questions

1. In which state are the eight balls at the lower energy (enthalpy)—state (i) or state (ii)?

2. a) Is ΔH positive or negative for the process: state (i) $\rightarrow$ state (ii)?

   b) Is the transformation from state (i) to state (ii) enthalpically favorable or unfavorable?

3. In which state are the eight balls at the higher entropy—state (i) or state (ii)?

4. a) Is ΔS positive or negative for the process: state (i) $\rightarrow$ state (ii)?

   b) Is the transformation from state (i) to state (ii) entropically favorable or unfavorable?

5. a) Is ΔG positive or negative for the process: state (i) $\rightarrow$ state (ii)?

   b) Is the transformation from state (i) to state (ii) thermodynamically favorable or unfavorable?

   c) Based on this analysis, is the transformation state (i) $\rightarrow$ state (ii) expected to be a naturally occurring process?

6. We place the eight balls in the left-side box, as in state (i). We wait several hours, but the transformation to state (ii) does not take place. Why? What needs to be done in order for the transformation state(i) $\rightarrow$ state(ii) to take place?

Model 2: A Theory of Reaction Rates.

The $\Delta G°$ for the reaction of dihydrogen and dioxygen (25°C) is very negative:

$$H_2(g) + \frac{1}{2}O_2(g) = H_2O(l) \qquad \Delta G° = -237 \text{ kJ/mol}_{rxn}$$

Thermodynamically, this reaction is expected to be a naturally occurring process (as is the process in Model 1). In fact, dihydrogen and dioxygen can be mixed at room temperature and no water is detected after months or years. However, if additional energy is provided (a spark, for example) the reaction does occur (explosively).

Many other chemical reactions are represented well by Model 1—they are thermodynamically favorable, but the reaction rate is exceptionally slow. A simple model has been proposed that attempts to explain the large variation in the observed rates of chemical reactions. This theory of reaction rates provides a basis for understanding why some chemical reactions are fast and others are slow:

- Molecules must collide in order for a reaction to occur.
- The rate at which molecules collide (the **collision frequency**) is generally greater than the rate of reaction involving those molecules. This suggests that not all collisions between molecules are effective in producing a reaction.
- There is a minimum energy of collision required for a reaction to occur. This is related to the fact that in order for the reaction to proceed, bonds must be broken. The state of the reacting system in which the molecules are colliding, and in which bonds are being broken (and new bonds are being formed), is known as the **transition state**, or the **activated complex**. The minimum energy needed to create this transition state is the **activation energy**, E_a.

The energetics of a reaction are often depicted by a diagram showing how the energy (enthalpy) of the molecules changes as the reaction proceeds—a **reaction coordinate diagram.** An example of such a diagram is shown below for the reaction $ONBr(g) + ONBr(g) \rightleftarrows 2NO(g) + Br_2(g)$.

Figure 1. Enthalpy versus Reaction Coordinate for a typical reaction.

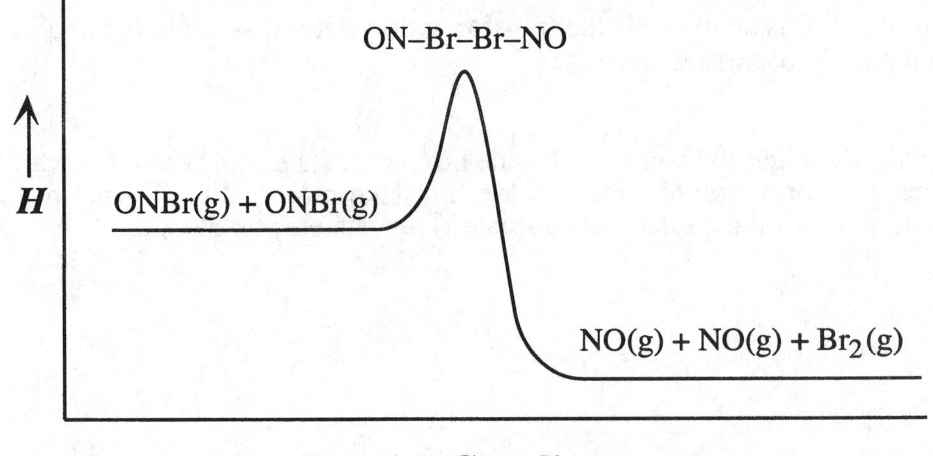

The reaction coordinate is a measure of the progress of a reaction. It represents all the changes that must occur in the course of the reaction, including the bending, breaking, and making of bonds.

Critical Thinking Questions

7. Which is more likely to weaken or break the bond between the nitrogen atom and the bromine atom in ONBr—a violent collision between two ONBr molecules or a gentle collision between two ONBr molecules? Explain.

8. Is a violent collision between two ONBr molecules more likely to occur at a high temperature or at a low temperature?

9. If two ONBr molecules collide as diagrammed below, is the collision likely to lead to two NO molecules and one Br_2 molecule? Explain.

   BrNO ⟶ ⟵ ONBr

10. Propose a more effective collision (one that is more likely to lead to two NO molecules and one Br_2 molecule) between these two molecules.

11. Provide two reasons (at least) why all collisions between molecules might not be effective in causing a reaction.

12. a) On Figure 1, draw a vertical line that indicates the magnitude of ΔH for the reaction.

    b) Is ΔH positive or negative?

    c) Is the reaction in Figure 1 an exothermic or endothermic reaction? Explain your reasoning.

13. Circle the activated complex (or transition state) in Figure 1.

14. Indicate on Figure 1 the magnitude of E_a for the forward reaction
    $ONBr(g) + ONBr(g) \rightarrow 2NO(g) + Br_2(g)$.

15. Indicate on Figure 1 the magnitude of E_a for the reverse reaction
    $2NO(g) + Br_2(g) \rightarrow 2ONBr(g)$.

16. What is the mathematical relationship between E_a(forward), E_a(reverse), and ΔH?

Exercises

1. Construct a reaction coordinate diagram (as in Figure 1) for a typical endothermic reaction. Which has the larger activation energy, the forward reaction or the reverse reaction?

2. The reaction $O_3(g) + NO(g) \rightarrow O_2(g) + NO_2(g)$ has $E_a = 10.7$ kJ/mole and $\Delta H = -199.8$ kJ/mole. What is the activation energy for the following reaction?

$$O_2(g) + NO_2(g) \rightarrow O_3(g) + NO(g)$$

3. The reaction $2\ N_2O(g) \rightleftarrows 2\ N_2(g) + O_2(g)$ is exothermic.

    a) Construct a reaction coordinate diagram (with enthalpy, H, on the y-axis and the reaction coordinate on the x-axis) for this reaction.
    b) Clearly indicate the magnitude of the activation energy for both the forward reaction and the reverse reaction with lines on the diagram.
    c) Give two reasons why not all collisions between N_2O molecules will necessarily be effective in causing this reaction to occur.

4. J. N. Spencer, G. M. Bodner, and L. H. Rickard, *Chemistry: Structure & Dynamics*, Third Edition, John Wiley & Sons, 2006. Chapter 14: Problem: 87.

Problems

1. The reaction $H_2O + H^+ \rightleftarrows H_3O^+$ has a very small activation energy (forward). Draw the Lewis structure for H_2O and give two reasons why the activation energy is small.

2. Why is the forward activation energy for the following reaction so large?

$$N_2(g) + 3H_2(g) \rightleftarrows 2NH_3(g)$$

ChemActivity 60

Reaction Mechanisms (II)

Chemical reactions occur on the molecular level by a sequence of one or more steps known as a **mechanism**. Every step is either a **unimolecular** or a **bimolecular** event. Typically, a single-headed arrow (→) is used to indicate a unimolecular or bimolecular event.

- A Unimolecular Step. A molecule undergoes a decomposition whereby a bond breaks and the molecule becomes two molecular fragments. The rate of a unimolecular step depends on the number of molecules present or, more simply, the concentration of the species.

Two examples of unimolecular events are shown below:

$Br_2(g) \rightarrow Br(g) + Br(g)$ rate = k' (Br_2)
$(CH_3)_3CBr(aq) \rightarrow (CH_3)_3C^+(aq) + Br^-(aq)$ rate = k'' ($(CH_3)_3CBr$)

- A Bimolecular Step. Two molecules collide, one or more bonds are broken, new bonds may or may not form. The rate of the step depends on the number of collisions between the two molecules (remember that not all collisions lead to a reaction). The number of collisions depends on the product of the concentrations of the two species.

Three examples of bimolecular events are shown below:

$CH_3Cl(aq) + I^-(aq) \rightarrow CH_3I(aq) + Cl^-(aq)$ rate = k''' (CH_3Cl) (I^-)
$CH_3I(aq) + Cl^-(aq) \rightarrow CH_3Cl(aq) + I^-(aq)$ rate = k'''' (CH_3I) (Cl^-)
$Br_2(g) + H(g) \rightarrow HBr(g) + Br(g)$ rate = k''''' (Br_2) (H)

The "k" for a bimolecular or unimolecular step is called a specific rate constant; the value of a specific rate constant depends on the molecular event specified and the temperature.

Note that the first two bimolecular events above are simply the reverse of each other. In this case, k''' = 5.2×10^{-7}/Ms and k'''' = 1.5×10^{-11}/Ms.

$$\text{forward rate} = \frac{5.2 \times 10^{-7}}{\text{Ms}} (CH_3Cl)(I^-)$$

$$\text{reverse rate} = \frac{1.5 \times 10^{-11}}{\text{Ms}} (CH_3I)(Cl^-)$$

These steps can be combined using a double arrow, and the specific rate constants can be used to calculate an equilibrium constant for the steps:

$$CH_3Cl(aq) + I^-(aq) \rightleftarrows CH_3I(aq) + Cl^-(aq) \qquad K = 3.5 \times 10^4$$

Every step in a mechanism is reversible. Often, however, the reverse reaction is negligible and not explicitly included in the mechanism.

The rate law for a proposed mechanism is determined by the sequence of steps that comprise the mechanism. However, the rate law is dominated by the slowest step, called the **rate-limiting step**.

Model 1: A Proposed Three-Step Mechanism for a Chemical Reaction.

Overall reaction: $(CH_3)_3CBr(aq) + OH^-(aq) \rightleftarrows (CH_3)_3COH(aq) + Br^-(aq)$

It is found experimentally that when the initial concentration of $(CH_3)_3CBr$ is doubled (keeping the initial hydroxide concentration constant) the rate of the reaction doubles. Furthermore, it is found that when the initial concentration of hydroxide is doubled (keeping the initial concentration of $(CH_3)_3CBr$ constant) the rate of the reaction remains the same.

Experimental rate law: rate = k_{exp} $((CH_3)_3CBr)$ $(OH^-)^o$ = k_{exp} $((CH_3)_3CBr)$

Proposed Mechanism:

Step	Molecular Event	Rate of Forward Step	Relative Rate
1	$(CH_3)_3CBr(aq) \rightleftarrows (CH_3)_3C^+(aq) + Br^-(aq)$	$k_1((CH_3)_3CBr)$	slow forward
2	$(CH_3)_3C^+(aq) + H_2O \rightleftarrows (CH_3)_3COH_2^+(aq)$	$k_2((CH_3)_3C^+)(H_2O)$	fast equilibrium
3	$(CH_3)_3COH_2^+(aq) + OH^-(aq) \rightleftarrows (CH_3)_3COH(aq) + H_2O$	$k_3((CH_3)_3COH_2^+)(OH^-)$	fast equilibrium

Rate law for proposed mechanism ≈ rate for step 1 = k_1 $((CH_3)_3CBr)$

Critical Thinking Questions

1. For this mechanism, which forward steps are unimolecular and which forward steps are bimolecular?

2. How was the rate law for the overall reaction determined?

3. Based on the information in Model 1 and the unimolecular and bimolecular examples given, how is the rate of a forward step determined from the molecular event?

4. Given the stoichiometry of the reaction, why isn't the rate law as follows?

   rate = $k_{experimental}$ $((CH_3)_3CBr)$ (OH^-)

5. Show that the sum of the three steps in the mechanism gives the stoichiometry of the overall reaction.

6. Why is the rate law for the proposed mechanism approximately equal to the rate of step 1?

7. Is the rate law for the proposed mechanism consistent with the experimental rate law? If not, why not?

Model 2: A Proposed Two-Step Mechanism for a Chemical Reaction.

Overall reaction: $2\ NO(g) + O_2(g) \rightleftarrows 2\ NO_2(g)$

Experimental rate law: rate = $k_{exp}\ (NO)^2\ (O_2)$

Proposed Mechanism:

Step	Molecular Event	Rate of Forward Step	Relative Rate
1	$NO(g) + NO(g) \rightleftarrows N_2O_2(g)$	$k_1\ (NO)^2$	fast equilibrium
2	$N_2O_2(g) + O_2(g) \rightleftarrows 2\ NO_2(g)$	$k_2\ (N_2O_2)\ (O_2)$	slow forward

Rate law for the proposed mechanism ≈ rate for step 2 = $k_2\ (N_2O_2)\ (O_2)$

Note that the species N_2O_2 does not appear in the overall reaction. Normally, it is quite difficult to measure the concentration of a reactive species that is not one of the reactants or products—called an **intermediate species**. For this reason, intermediate species are not normally included in rate laws. In this mechanism, note that step 1 is fast, contains N_2O_2 and NO (a reactant), and is at equilibrium. Step 1 should remain in an equilibrium state as step 2 slowly consumes O_2. Thus,

Step 1 : $NO + NO \rightleftarrows N_2O_2$ fast equilibrium

rate of forward step = rate of reverse step

$$k_1\ (NO)^2 = k_{-1}\ (N_2O_2)$$

Thus, $(N_2O_2) = \frac{k_1}{k_{-1}} (NO)^2 = K\ (NO)^2$, where K is the equilibrium constant for step 1.

Rate law for the proposed mechanism ≈ rate for step 2 = $k_2\ (N_2O_2)\ (O_2)$

$$= k_2\ K\ (NO)^2\ (O_2) = k_{exp}\ (NO)^2\ (O_2)$$

Note that it is not necessary to know the values of k_2 or K to show that this mechanism leads to a rate law that is consistent with experimental data. However, to prove that this mechanism is the correct mechanism, it might be necessary to experimentally determine k_2 and K and to show that $k_2 \times K = k_{exp}$.

Critical Thinking Questions

8. For this mechanism, which forward steps are unimolecular and which forward steps are bimolecular?

9. Is the "rate of forward step" given for each step consistent with your answer to CTQ 3?

10. Why is the rate law for the proposed mechanism approximately equal to the rate of forward step 2?

11. Is the rate law for the proposed mechanism consistent with the experimental rate law for the overall reaction?

12. Show that the sum of the two steps in the mechanism gives the stoichiometry of the overall reaction.

Exercises

1. Indicate the molecularity (unimolecular or bimolecular) of each of the following steps. Give the rate for each step (the first process is shown as an example):

   $O_3(g) \rightarrow O_2(g) + O(g)$ unimolecular rate = k' (O_3)
   $ONBr(g) + ONBr(g) \rightarrow NO(g) + NO(g) + Br_2(g)$
   $N_2O_2(g) \rightarrow NO(g) + NO(g)$
   $NO(g) + NO(g) \rightarrow N_2(g) + O_2(g)$
   $I(g) + H_2(g) \rightarrow HI(g) + H(g)$

2. The following reaction is first order with respect to both NO and F_2:[1]

   $2\ NO_2(g) + F_2(g) \rightleftarrows 2\ NO_2F(g)$

   rate = $k_{exp}\ (NO_2)\ (F_2)$

   This rate law is consistent with which of the following mechanisms?

   a) $NO_2 + F_2 \rightleftarrows NO_2F + F$ fast
   $NO_2 + F \rightleftarrows NO_2F$ slow

   b) $NO_2 + F_2 \rightleftarrows NO_2F + F$ slow
   $NO_2 + F \rightleftarrows NO_2F$ fast

   c) $F_2 \rightleftarrows F + F$ slow
   $2\ NO_2 + 2\ F \rightleftarrows 2\ NO_2F$ fast

   Add the molecular species for the two steps in each of the mechanisms. How is this sum related to the stoichiometry of the overall reaction?

3. J. N. Spencer, G. M. Bodner, and L. H. Rickard, *Chemistry: Structure & Dynamics*, Third Edition, John Wiley & Sons, 2006. Chapter 14: Problems: 37-39.

---

[1] J. N. Spencer, G. M. Bodner, and L. H. Rickard, *Chemistry: Structure & Dynamics*, Third Edition, John Wiley & Sons, 2006, Chapter 14, Problem 35.

Model 3: Thermodynamic and Kinetic Control.

Many chemical reactions are thermodynamically unfavorable (K<1). These reactions may occur to a very limited extent (as determined from $\Delta G°$ and the equilibrium constant) . They are said to be under **thermodynamic control**:

$$CH_3COOH(aq) \rightleftarrows CH_3COO^-(aq) + H^+(aq) \qquad K_a = 1.8 \times 10^{-5}$$

$$2\,Cl^-(1M) + Br_2(l) \rightleftarrows 2\,Br^-(1M) + Cl_2(g) \qquad K = 3.9 \times 10^{-11}$$

Many chemical reactions are thermodynamically favorable and proceed to virtual completion:

$$2\,Na(s) + 2\,H_2O(l) \rightleftarrows 2\,Na^+(aq) + 2\,OH^-(aq) + H_2(g) \qquad K = 10^{77}$$

$$Zn(s) + Cu^{2+}(aq) \rightleftarrows Cu(s) + Zn^{2+}(aq) \qquad K = 10^{37}$$

Some chemical reactions are thermodynamically favorable but no reaction is apparent over long time periods.

$$H_2(g) + \frac{1}{2}O_2(g) \rightleftarrows H_2O(l) \qquad K = 10^{41}$$

$$C(diamond) \rightleftarrows C(graphite) \qquad K = 3.2$$

$$C(diamond) + O_2(g) \rightleftarrows CO_2(g) \qquad K = 10^{69}$$

These reactions are said to be under **kinetic control.**

Critical Thinking Questions

13. For a certain chemical reaction $\Delta G° = 200$ kJ/mol. When the reactants are mixed, no chemical reaction is apparent. Is this reaction under thermodynamic or kinetic control?

14. For a certain chemical reaction $\Delta G° = -200$ kJ/mol. When the reactants are mixed, no chemical reaction is apparent. Is this reaction under thermodynamic or kinetic control?

15. What one feature of a reaction coordinate diagram is indicative of kinetic control of a reaction? Explain.

    a) $\Delta H° < 0$

    b) $\Delta H° > 0$

    c) Activation energy is large

    d) Activation energy is small

Exercises

4. The equilibrium constant for the following reaction is quite large. Is it possible to predict the extent of reaction in a reasonable time period (a few minutes or hours)?

   $Zn(s) + 2\ H^+(aq) \rightleftarrows Zn^{2+}(aq) + H_2(g)$

5. The equilibrium constant for the following reaction is quite small. Is it possible to predict the extent of reaction in a reasonable time period (a few minutes or hours)?

   $Cu(s) + 2\ H^+(aq) \rightleftarrows Cu^{2+}(aq) + H_2(g)$

6. The diamond to graphite reaction is thermodynamically favorable but does not appear to happen during the lifetime of an engagement ring. Use the appropriate tables to determine if this reaction is endothermic or exothermic. Construct a reaction coordinate diagram that shows the endothermic or exothermic nature of the reaction and illustrates why this reaction is under kinetic control.

7. Each diagram below (I, II, III, IV) describes a possible reaction:

   $A_2(g) + B_2(g) \rightleftarrows 2\ AB(g)$

   Assuming that you begin with equal amounts of $A_2(g)$ and $B_2(g)$, but no AB, and assuming that $\Delta S°$ is the same for all of the possible reactions, for which of these diagrams would:

   a) the reaction proceed fastest in the forward direction?
   b) the amount of AB(g) at equilibrium be the greatest?
   c) the equilibrium constant be the smallest?
   d) equilibrium be reached in the shortest amount of time?

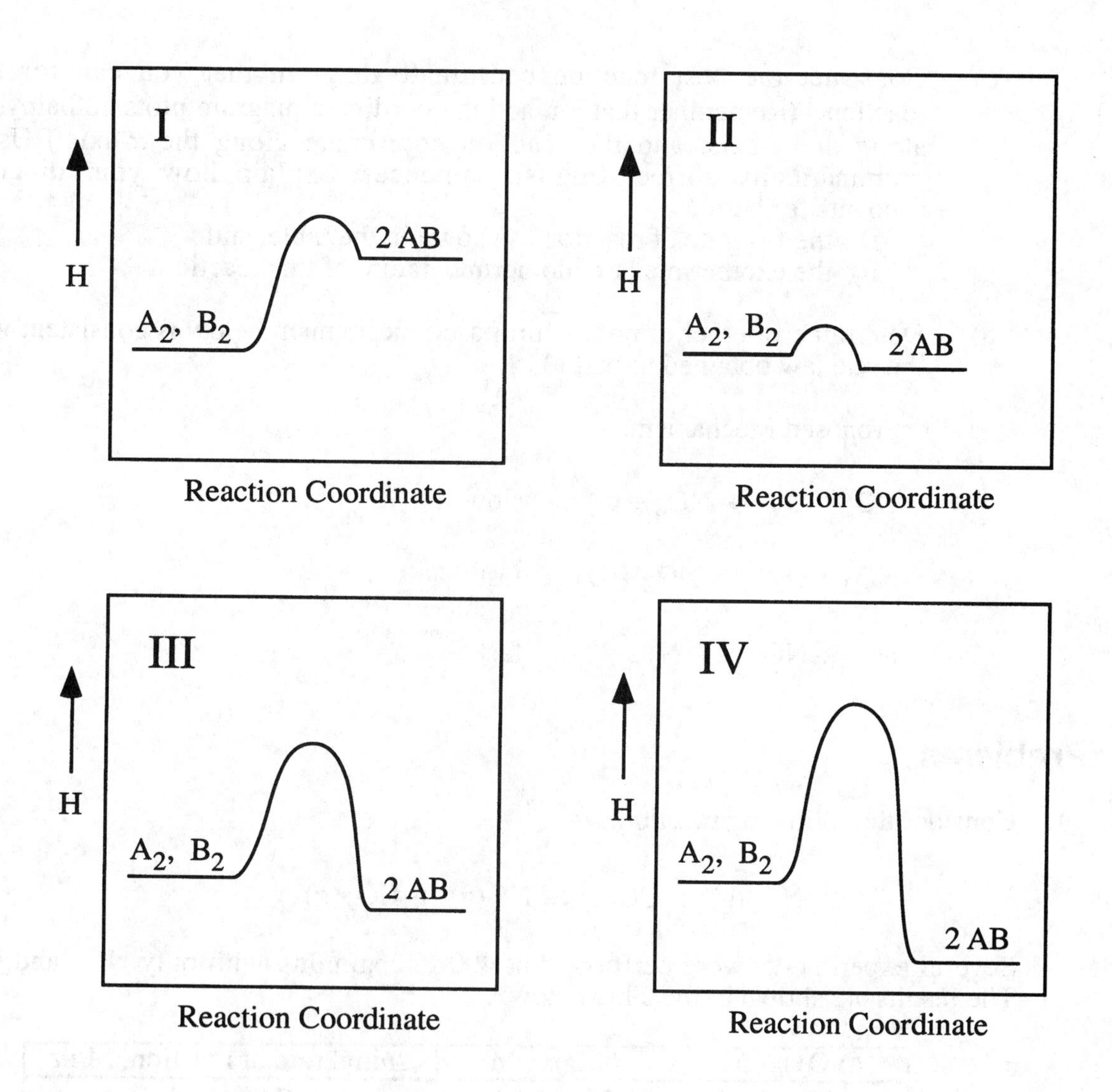

8. Consider the following reaction:

$$2\ NO_2(g) + O_3(g) \rightleftarrows N_2O_5(g) + O_2(g)$$

Several experiments were performed at 298 K beginning with only NO_2 and O_3. The results are shown in the table below:

$(NO_2)_o$, M	$(O_3)_o$, M	Initial rate of reaction, M/sec
0.50	1.00	2.5×10^4
2.00	1.00	1.0×10^5
2.00	2.00	2.0×10^5

a) Determine the rate law for this reaction, and the value of the rate constant at 298 K.

b) Determine $\Delta G°$ at 298 K for this reaction.

c) Construct the best reaction coordinate diagram that you can for this reaction. (Remember that a reaction coordinate diagram plots enthalpy, *H*, along the *y*-axis, and the reaction coordinate along the *x*-axis.) Using grammatically correct English sentences, explain how your diagram accounts for both
   i) the fast rate of reaction as shown in the table, and
   ii) the exothermic or endothermic nature of this reaction.

d) Determine whether or not the proposed mechanism below is consistent with the rate law obtained in part a).

Proposed Mechanism:

$NO_2 + O_3 \rightarrow NO_3 + O_2$ slow

$NO_3 + O_2 \rightarrow NO_2 + O_3$ fast

$NO_3 + NO_2 \rightarrow N_2O_5$ fast

Problems

1. Consider the following reaction:

$$NO_2(g) + CO(g) \rightleftarrows NO(g) + CO_2(g)$$

Several experiments were performed at 400 K beginning with only NO_2 and CO. The results are shown in the table below:

$(NO_2)_o$, M	$(CO)_o$, M	Initial rate of reaction, M/hr
0.38×10^{-4}	5.1×10^{-4}	3.5×10^{-8}
0.76×10^{-4}	5.1×10^{-4}	1.4×10^{-7}
0.38×10^{-4}	8.6×10^{-4}	3.5×10^{-8}

a) Determine the rate law for this reaction, and the value of the rate constant at 400 K.

b) Determine whether or not the proposed mechanism below is consistent with the rate law obtained in part a).

PROPOSED MECHANISM:

$NO_2 + NO_2 \rightarrow NO_3 + NO$ slow

$NO_3 + CO \rightarrow NO_2 + CO_2$ fast

2. The conversion of ozone to molecular oxygen in the upper atmosphere,

$$2\,O_3(g) \rightleftarrows 3\,O_2(g)$$

is thought to occur via the following mechanism:

$O_3 \rightleftarrows O_2 + O$ (fast equilibrium)
$O + O_3 \rightleftarrows 2\,O_2$ (slow forward)

a) What is the rate law for this mechanism (remember that only the concentrations of the reactants and the products can appear in the rate law)? b) The rate law from the mechanism above is consistent with the experimental rate law. Explain the experimental fact that the rate decreases as the concentration of O_2 increases.

3. A possible mechanism for a chemical reaction is:

$Fe^{2+}(aq) + I_2(aq) \rightleftarrows Fe^{3+}(aq) + I_2^-(aq)$ (fast equilibrium)

$Fe^{2+}(aq) + I_2^-(aq) \rightarrow Fe^{3+}(aq) + 2\,I^-(aq)$ (slow forward)

What is the overall chemical reaction for this mechanism?

ChemActivity 61

Reaction Mechanisms (III)

(What Is a Catalyst?)

Model 1: A Catalyst.

A **catalyst** is a substance that is neither produced nor consumed in a chemical reaction, yet causes the rate of the reaction to be increased without changing the temperature. As an example, the reaction $H_2(g) + I_2(g) \rightleftarrows 2\ HI(g)$ proceeds about 10^8 times faster in the presence of Pt dust than without it. In this case, Pt acts as a catalyst. The presence of a catalyst enables a reaction to take place using a different mechanism than would otherwise be possible.

Critical Thinking Questions

1. What effect does a catalyst have on the stoichiometry of the balanced chemical equation describing a reaction? Explain your reasoning.

2. What effect does a catalyst have on the $\Delta H°$ of a chemical reaction? Explain your reasoning.

3. How does the rate of the rate-limiting step in a mechanism involving a catalyst compare to the rate of the rate-limiting step of the mechanism without the catalyst present? Explain your reasoning.

4. What effect does a catalyst have on the activation energy of an overall reaction?

Model 2: Hydrolysis of Glycylglycine.

In the following reaction glycylglycine is split into two glycine molecules by a water molecule.

$$H_2NCH_2CONHCH_2COOH(aq) + H_2O(l) \rightleftarrows 2\ H_2NCH_2COOH(aq) \qquad (1)$$

The mechanism is one-step bimolecular process wherein the partial negatively charged oxygen atom of the water molecule collides with the partial positively charged carbon atom, as shown below.

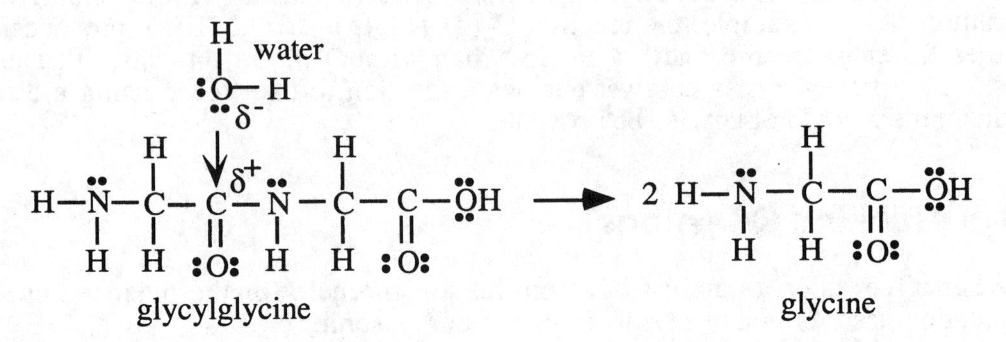

The reaction is thermodynamically favorable, but the activation energy is very high and the reaction is extremely slow.

The hydrolysis of glycylglycine, however, is extremely rapid in the presence of a catalyst. Biological catalysts are called **enzymes**. In this case the enzyme is a large protein containing a Co^{2+} ion that can bond to six atoms (an octahedral arrangement).

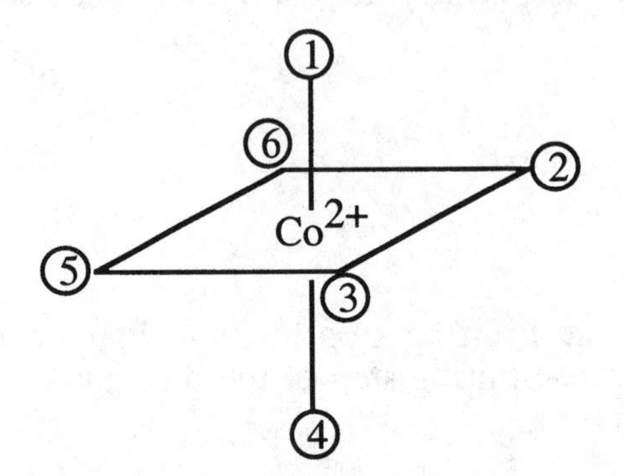

Three of the sites are occupied by nitrogen atoms in the protein—sites 4, 5, and 6 in the diagram above. The other three sites *can* be occupied by two nitrogen atoms and one oxygen atom of glycylglycine. The cavity in the enzyme is sufficiently large to permit glycylglycine and other small ions (Cl^-) and molecules (H_2O) to bind to the cobalt, but larger molecules are physically excluded. This enzyme has but one purpose—to facilitate the conversion of glycylglycine to glycine. The collision between water and the partial positively charged carbon atom is more successful when the glycylglycine is bonded to the enzyme because the Co^{2+} ion pulls electrons away from the glycylglycine and increases the partial positive charge on the carbon atom. Two glycine molecules do not fit very well into the cavity. The glycine molecules leave the cavity, and the enzyme is ready to accept another glycylglycine molecule.

Figure 1. The glycylglycine molecule bonds to the enzyme.

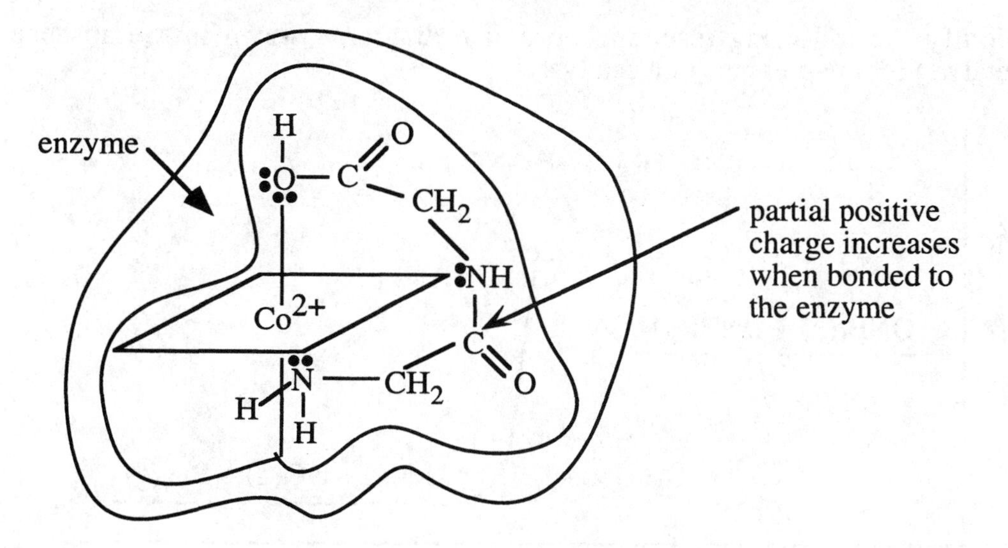

5. Is the reaction in the model a balanced chemical equation?

6. Where do the two hydrogen atoms and the oxygen atom from the water molecule end up?

7. a) Is reaction (1) a redox reaction? How can you tell?

   b) Is reaction (1) an acid/base reaction? How can you tell?

8. Is the uncatalyzed reaction under thermodynamic control or kinetic control? Explain.

9. In Figure 1, why does the Co^{2+} ion pull electrons away from glycylglycine?

Exercises

1. Modify the following reaction coordinate diagram (drawn in the absence of a catalyst) for the presence of a catalyst.

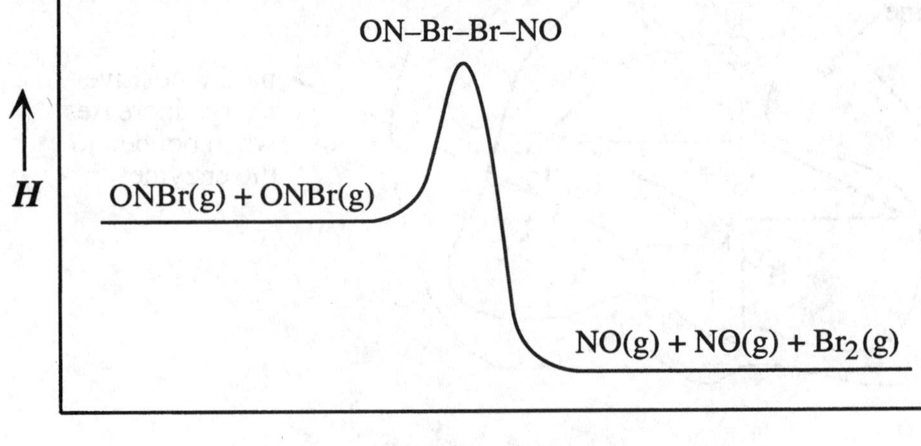

2. Assume that the glycylglycine to glycine reaction is exothermic. Sketch the reaction coordinate diagram for this reaction. Modify the diagram for the presence of the enzyme.

3. J. N. Spencer, G. M. Bodner, and L. H. Rickard, *Chemistry: Structure & Dynamics*, Third Edition, John Wiley & Sons, 2006. Chapter 14: Problems: 88, 92, 99, 100.

Problem

1. At 1000°C, the reaction $2\,HI(g) \rightleftarrows I_2(g) + H_2(g)$ has E_a (forward) = 183 kJ/mol and $\Delta H° = 9.5$ kJ/mole. a) Draw a reaction coordinate diagram (H vs. reaction coordinate) for this reaction. Graph paper is not necessary, but make some attempt to scale properly. Clearly indicate E_a (forward), E_a (reverse), and $\Delta H°$on the diagram. b) Determine the value of E_a (reverse). c) A platinum catalyst is added and E_a (forward) is reduced to 58 kJ/mole. What is the value of $\Delta H°$ when the catalyst is present?

ChemActivity 62

Activation Energy

Model: The Arrhenius Equation

The rate of a reaction depends on the temperature. This occurs because the value of the rate constant, k, often is a function of the absolute temperature, T, measured in Kelvin. In general, the relationship between k and T is found to be

$$\ln \mathrm{k} = \ln A - \frac{E_a}{RT} \qquad (1)$$

Here, A is the **frequency factor**, E_a is the **activation energy** in units of joules per mole, and R is the gas constant (8.314 J/mol K). Both A and E_a are characteristic of the particular reaction being studied. If the rate constant for a given reaction is examined at two temperatures, T_1 and T_2, and if the observed rate constants at those temperatures are k_1 and k_2, respectively, then equation 1 can be used to derive the following relationship:

$$\ln \frac{k_1}{k_2} = \frac{E_a}{R}\left[\frac{1}{T_2} - \frac{1}{T_1}\right] \qquad (2)$$

Critical Thinking Questions

1. Based on equation (1), what is the slope and what is the intercept of a plot of ln k versus $1/T$?

2. According to equation (1), if the activation energy for some reaction, Q, is greater than the activation energy for a different reaction, W, which has the greater rate constant—reaction Q or reaction W (assuming that the value of A, the frequency factor, is identical in both reactions)? Explain.

3. According to equation (1), if the temperature increases, does the rate constant increase or decrease? Explain.

4. Give an explanation for the effect of temperature on the rate constant based on molecular speeds.

5. At how many different temperatures must the rate constant be determined in order to evaluate the activation energy for a reaction?

Exercises

1. A chemist's "rule of thumb" is that the rate of a chemical reaction doubles for every 10°C increase in temperature. Use equation (2) to demonstrate this rule of thumb (assume that a typical chemical reaction has an activation energy of 50 kJ/mol). Recall that $R = 8.314 \frac{\text{J}}{\text{K mol}}$. (Hint: a typical chemical reaction occurs at a typical temperature.)

2. A great Martian chemist enunciated the following chemical principle:

   The rate of a chemical reaction doubles for every 7°C increase in temperature.

   Assume that the average temperature on Mars is –40°C, and determine if the Martian chemist was correct or not.

3. Indicate whether the following statement is true or false and explain your reasoning.

   In general, the higher the activation energy, the faster a reaction occurs at a given temperature.

4. J. N. Spencer, G. M. Bodner, and L. H. Rickard, *Chemistry: Structure & Dynamics*, Third Edition, John Wiley & Sons, 2006. Chapter 14: Problems: 93-97, 104, 105, 110, 111.

Problems

1. $H_2(g) + I_2(g) \rightarrow 2\,HI(g)$

The rate constant for the above reaction at two temperatures was determined:

Temperature (K)	Rate Constant ($M^{-1}\ sec^{-1}$)
400	0.0234
500	0.750

Determine the rate constant at 400 K for the reverse reaction

$$2\,HI(g) \rightarrow H_2(g) + I_2(g)$$

as precisely as you can, assuming that $A = 5 \times 10^5\ M^{-1}\ s^{-1}$ for the reverse reaction.

2. Consider a generic reaction:

$$AB\ (g)\ +\ CD\ (g) \rightleftarrows AC\ (g)\ +\ BD\ (g)$$

a) Construct a reaction coordinate diagram for this reaction assuming that it has a large equilibrium constant, but that it reaches equilibrium very slowly. Explain your reasoning clearly.

b) Indicate how the addition of a catalyst would change the reaction coordinate diagram from part a), and describe what effect this would have on the equilibrium constant and the rate at which equilibrium is reached.

Appendix

TABLE A.1 Values of Selected Fundamental Constants

Speed of light in a vacuum (c)	$c = 2.99792458 \times 10^{8}$ m/s
Charge on an electron (q_e)	$q_e = 1.6021892 \times 10^{-19}$ C
Rest mass of an electron (m_e)	$m_e = 9.109534 \times 10^{-28}$ g
	$m_e = 5.4858026 \times 10^{-4}$ amu
Rest mass of a proton (m_p)	$m_p = 1.6726485 \times 10^{-24}$ g
	$m_p = 1.00727647$ amu
Rest mass of a neutron (m_n)	$m_n = 1.6749543 \times 10^{-24}$ g
	$m_n = 1.008665012$ amu
Faraday's constant (F)	$F = 96{,}484.56$ C/mol
Planck's constant (h)	$h = 6.626176 \times 10^{-34}$ J · s
Ideal gas constant (R)	$R = 0.0820568$ L-atm/mol-K
	$R = 8.31441$ J/mol-K
Atomic mass unit (amu)	1 amu = $1.6605655 \times 10^{-24}$ g
Boltzmann's constant (k)	$k = 1.380662 \times 10^{-23}$ J/K
Avogadro's constant (N)	$N = 6.022045 \times 10^{23}$ mol^{-1}
Rydberg constant (R_H)	$R_H = 1.09737318 \times 10^{7}$ m^{-1}
	$= 1.09737318 \times 10^{-2}$ nm^{-1}
Heat capacity of water	$C = 75.376$ J/mol-K

TABLE A.2 Selected Conversion Factors

Energy	1 J = 0.2390 cal = 10^{7} erg = 1 volt•coulomb
	1 cal = 4.184 J (by definition)
	1 ev/atom = $1.6021892 \times 10^{-19}$ J/atom = 96.484 kJ/mol
Temperature	K = °C + 273.15
	°C = (5/9)(°F – 32)
	°F = (9/5)(°C) + 32
Pressure	1 atm = 760 mm Hg = 760 torr = 101.325 kPa
Mass	1 kg = 2.2046 lb
	1 lb = 453.59 g = 0.45359 kg
	1 oz = 0.06250 lb = 28.350 g
	1 ton = 2000 lb = 907.185 kg
	1 tonne (metric) = 1000 kg = 2204.62 lb
Volume	1 mL = 0.001 L = 1 cm^{3} (by definition)
	1 oz (fluid) = 0.031250 qt = 0.029573 L
	1 qt = 0.946326 L
	1 L = 1.05672 qt
Length	1 m = 39.370 in
	1 mi = 1.60934 km
	1 in = 2.54 cm (by definition)

TABLE A.3 Standard-State Enthalpies, Free Energies, and Entropies of Atom Combination

Substance	ΔH°_{ac} (kJ/mol)	ΔG°_{ac} (kJ/mol)	ΔS°_{ac} (J/mol-K)
	Aluminum		
$Al(s)$	–326.4	–285.7	–136.21
$Al(g)$	0	0	0
$Al^{+3}(aq)$	–857	–77.1	–486.2
$Al_2O_3(s)$	–3076.0	–2848.9	–761.33
$AlCl_3(s)$	–1395.6	–1231.5	–549.46
$AlF_3(s)$	–2067.5	–1896.4	–574.36
$Al_2(SO_4)_3(s)$	–7920.1	–7166.9	–2525.9
	Barium		
$Ba(s)$	–180	–146	–107.4
$Ba(g)$	0	0	0
$Ba^{+2}(aq)$	–718	–707	–160.6
$BaO(s)$	–983	–903	–260.88
$Ba(OH)_2 \cdot 8H_2O(s)$	–9931.6	–8915	–3419
$BaCl_2(s)$	–1282	–1168	–376.96
$BaCl_2(aq)$	–1295	–1181	–378.04
$BaSO_4(s)$	–2929	–2673	–850.1
$Ba(NO_3)_2(s)$	–3612	–3217	–1229.4
$Ba(NO_3)_2(aq)$	–3573	–3231	–1140.7
	Beryllium		
$Be(s)$	–324.3	–286.6	–126.77
$Be(g)$	0	0	0
$Be^{+2}(aq)$	–707.1	–666.3	–266.0
$BeO(s)$	–1183.1	–1026.6	–283.18
$BeCl_2(s)$	–1058.1	–943.6	–383.99
	Bismuth		
$Bi(s)$	–207.1	–168.2	–130.31
$Bi(g)$	0	0	0
$Bi_2O_3(s)$	–1735.6	–1525.3	–705.8
$BiCl_3(s)$	–951.2	–800.2	–505.6
$BiCl_3(g)$	–837.8	–741.2	–323.79
$Bi_2S_3(s)$	–1393.7	–1191.8	–677.2
	Boron		
$B(s)$	–562.7	–518.8	–147.59
$B(g)$	0	0	0
$B_2O_3(s)$	–3145.7	–2926.4	–736.10
$B_2H_6(g)$	–2395.7	–2170.4	–763.07
$B_5H_9(l)$	–4729.7	–4251.4	–1615.45
$B_{10}H_{14}(s)$	–8719.3	–7841.2	–2963.92

(continued)

TABLE A.3 Standard-State Enthalpies, Free Energies, and Entropies of Atom Combination (continued)

Substance	ΔH°_{ac} (kJ/mol)	ΔG°_{ac} (kJ/mol)	ΔS°_{ac} (J/mol-K)
H_3BO_3(s)	–3057.5	–2792.7	–891.92
BF_3(g)	–1936.7	–1824.9	–375.59
BCl_3(l)	–1354.9	–1223.2	–442.7
$B_3N_3H_6$(l)	–4953.1	–4535.4	–1408.9
$B_3N_3H_6$(g)	–4923.9	–4532.8	–1319.84
	Bromine		
Br_2(l)	–223.768	–164.792	–197.813
Br_2(g)	–192.86	–161.68	–104.58
Br(g)	0	0	0
HBr(g)	–365.93	–339.09	–91.040
HBr(aq)	–451.08	–389.60	–207.3
BrF(g)	–284.72	–253.49	–104.81
BrF_3(g)	–604.45	–497.56	–358.75
BrF_5(g)	–935.73	–742.6	–648.60
	Calcium		
Ca(s)	–178.2	–144.3	–113.46
Ca(g)	0	0	0
Ca^{+2}(aq)	–721.0	–697.9	–208.0
CaO(s)	–1062.5	–980.1	–276.19
$Ca(OH)_2$(s)	–2097.9	–1912.7	–623.03
$CaCl_2$(s)	–1217.4	–1103.8	–380.7
$CaSO_4$(s)	–2887.8	–2631.3	–860.3
$CaSO_4 \cdot 2H_2O$(s)	–4256.7	–4383.2	–1553.8
$Ca(NO_3)_2$(s)	–3557.0	–3189.0	–1234.5
$CaCO_3$(s)	–2849.3	–2639.5	–703.2
$Ca_3(PO_4)_2$(s)	–7278.0	–6727.9	–1843.5
	Carbon		
C(graphite)	–716.682	–671.257	–152.36
C(diamond)	–714.787	–668.357	–155.719
C(g)	0	0	0
CO(g)	–1076.377	–1040.156	–121.477
CO_2(g)	–1608.531	–1529.078	–266.47
$COCl_2$(g)	–1428.0	–1318.9	–366.02
CH_4(g)	–1662.09	–1535.00	–430.68
HCHO(g)	–1509.72	–1412.01	–329.81
H_2CO_3(aq)	–2599.14	–2396.02	–683.3
HCO_3^-(aq)	–2373.83	–2156.47	–664.8
CO_3^{-2}(aq)	–2141.33	–1894.26	–698.16
CH_3OH(l)	–2075.11	–1882.25	–651.2
CH_3OH(g)	–2037.11	–1877.94	–538.19
CCl_4(l)	–1338.84	–1159.19	–602.49
CCl_4(g)	–1306.3	–1154.57	–509.04
$CHCl_3$(l)	–1433.84	–1265.20	–567.3

TABLE A.3 Standard-State Enthalpies, Free Energies, and Entropies of Atom Combination (continued)

Substance	ΔH°_{ac} (kJ/mol)	ΔG°_{ac} (kJ/mol)	ΔS°_{ac} (J/mol-K)
$CHCl_3(g)$	–1402.51	–1261.88	–472.69
$CH_2Cl_2(l)$	–1516.80	–1356.37	–540.1
$CH_2Cl_2(g)$	–1487.81	–1354.98	–447.69
$CH_3Cl(g)$	–1572.15	–1444.08	–432.9
$CS_2(l)$	–1184.59	–1082.49	–342.40
$CS_2(g)$	–1156.93	–1080.64	–255.90
$HCN(g)$	–1271.9	–1205.43	–224.33
$CH_3NO_2(l)$	–2453.77	–2214.51	–805.89
$C_2H_2(g)$	–1641.93	–1539.81	–344.68
$C_2H_4(g)$	–2251.70	–2087.35	–555.48
$C_2H_6(g)$	–2823.94	–2594.82	–774.87
$CH_3CHO(l)$	–2745.43	–2515.35	–775.9
$CH_3CO_2H(l)$	–3286.8	–3008.86	–937.4
$CH_3CO_2H(g)$	–3234.55	–2992.96	–814.7
$CH_3CO_2H(aq)$	–3288.06	–3015.42	–918.5
$CH_3CO_2^-(aq)$	–3070.66	–2785.03	–895.8
$CH_3CH_2OH(l)$	–3266.12	–2968.51	–1004.8
$CH_3CH_2OH(g)$	–3223.53	–2962.22	–882.82
$CH_3CH_2OH(aq)$	–3276.7	–2975.37	–1017.0
$C_6H_6(l)$	–5556.96	–5122.52	–1464.1
$C_6H_6(g)$	–5523.07	–5117.36	–1367.7
	Chlorine		
$Cl_2(g)$	–243.358	–211.360	–107.330
$Cl(g)$	0	0	0
$Cl^-(aq)$	–288.838	–236.908	–108.7
$ClO_2(g)$	–517.5	–448.6	–230.47
$Cl_2O(g)$	–412.2	–345.2	–225.24
$Cl_2O_7(l)$	–1750	-	-
$HCl(g)$	–431.64	–404.226	–93.003
$HCl(aq)$	–506.49	–440.155	–223.4
$ClF(g)$	–255.15	–223.53	–106.06
	Chromium		
$Cr(s)$	–396.6	–351.8	–150.73
$Cr(g)$	0	0	0
$CrO_3(s)$	–1733.6	-	-
$CrO_4^{-2}(aq)$	–2274.4	–2006.47	–768.51
$Cr_2O_3(s)$	–2680.4	–2456.89	–751.0
$Cr_2O_7^{-2}(aq)$	–4027.7	–3626.8	–1214.5
$(NH_4)_2Cr_2O_7$	–7030.7	-	-
$PbCrO_4(s)$	–2519.2	-	-
	Cobalt		
$Co(s)$	–424.7	–380.3	–149.475
$Co(g)$	0	0	0

(continued)

TABLE A.3 Standard-State Enthalpies, Free Energies, and Entropies of Atom Combination (continued)

Substance	ΔH°_{ac} (kJ/mol)	ΔG°_{ac} (kJ/mol)	ΔS°_{ac} (J/mol-K)
$Co^{+2}(aq)$	–482.9	–434.7	–293
$Co^{+3}(aq)$	–333	–246.3	–485
$CoO(s)$	–911.8	–826.2	–287.60
$Co_3O_4(s)$	–3162	–2842	–1080.3
$Co(NH_3)_6{}^{+3}(aq)$	–7763.5	–6929.5	–3018.
	Copper		
$Cu(s)$	–338.32	–298.58	–133.23
$Cu(g)$	0	0	0
$Cu^{+}(aq)$	–266.65	–248.60	–125.8
$Cu^{+2}(aq)$	–273.55	–233.09	–266.0
$CuO(s)$	–744.8	–660.0	–284.81
$Cu_2O(s)$	–1094.4	–974.9	–400.68
$CuCl_2(s)$	–807.8	–685.6	–388.71
$CuS(s)$	–670.2	–590.4	–267.7
$Cu_2S(s)$	–1304.9	–921.6	–379.7
$CuSO_4(s)$	–2385.17	–1530.44	–869
$Cu(NH_3)_4{}^{+2}(aq)$	–5189.4	–4671.13	–1882.5
	Fluorine		
$F_2(g)$	–157.98	–123.82	–114.73
$F(g)$	0	0	0
$F^-(aq)$	–411.62	–340.70	–172.6
$HF(g)$	–567.7	–538.4	–99.688
$HF(aq)$	–616.72	–561.98	–184.8
	Hydrogen		
$H_2(g)$	–435.30	–406.494	–98.742
$H(g)$	0	0	0
$H^{+}(aq)$	–217.65	–203.247	–114.713
$OH^-(aq)$	–696.81	–592.222	–286.52
$H_2O(l)$	–970.30	–875.354	–320.57
$H_2O(g)$	–926.29	–866.797	–202.23
$H_2O_2(l)$	–1121.42	–990.31	–441.9
$H_2O_2(aq)$	–1124.81	–1003.99	–407.6
	Iodine		
$I_2(s)$	–213.676	–141.00	–245.447
$I_2(g)$	–151.238	–121.67	–100.89
$I(g)$	0	0	0
$HI(g)$	–298.01	–272.05	–88.910
$IF(g)$	–281.48	–250.92	–103.38
$IF_5(g)$	–1324.28	–1131.78	–646.9
$IF_7(g)$	–1603.7	–1322.17	–945.6
$ICl(g)$	–210.74	–181.64	–98.438
$IBr(g)$	–177.88	–149.21	–97.040

TABLE A.3 Standard-State Enthalpies, Free Energies, and Entropies of Atom Combination (continued)

Substance	ΔH°_{ac} (kJ/mol)	ΔG°_{ac} (kJ/mol)	ΔS°_{ac} (J/mol-K)
	Iron		
Fe(s)	–416.3	–370.7	–153.21
Fe(g)	0	0	0
Fe^{+2}(aq)	–505.4	–449.6	–318.2
Fe^{+3}(aq)	–464.8	–375.4	–496.4
Fe_2O_3(s)	–2404.3	–2178.8	–756.75
Fe_3O_4(s)	–3364.0	–3054.4	–1039.3
$Fe(OH)_2$(s)	–1918.9	–1727.2	–644
$Fe(OH)_3$(s)	–2639.8	–2372.1	–901.1
$FeCl_3$(s)	–1180.8	–1021.7	–533.8
FeS_2(s)	–1152.1	–1014.1	–463.2
$Fe(CO)_5$(l)	–6019.6	–5590.9	–1438.1
$Fe(CO)_5$(g)	–5979.5	–5582.9	–1330.9
	Lead		
Pb(s)	–195.0	–161.9	–110.56
Pb(g)	0	0	0
Pb^{+2}(aq)	–196.7	–186.3	–164.9
PbO(s)	–661.5	–581.5	–267.7
PbO_2(s)	–970.7	–842.7	–428.9
$PbCl_2$(s)	–797.8	–687.4	–369.8
$PbCl_4$(l)	–1011.0	-	-
PbS(s)	–574.2	–498.9	–252.0
$PbSO_4$(s)	–2390.4	–2140.2	–838.84
$Pb(NO_3)_2$(s)	–3087.3	-	-
$PbCO_3$(s)	–2358.3	–2153.8	–685.6
	Lithium		
Li(s)	–159.37	–126.66	–109.65
Li(g)	0	0	0
Li^{+}(aq)	–437.86	–419.97	–125.4
LiH(s)	–467.56	–398.26	–233.40
LiOH(s)	–1111.12	–1000.59	–371.74
LiF(s)	–854.33	–784.28	–261.87
LiCl(s)	–689.66	–616.71	–244.64
LiBr(s)	–622.48	–551.06	–239.52
LiI(s)	–536.62	–467.45	–232.78
$LiAlH_4$(s)	–1472.7	–1270.0	–683.42
$LiBH_4$(s)	–1401.9	–1333.4	–675.21
	Magnesium		
Mg(s)	–147.70	–113.10	–115.97
Mg(g)	0	0	0
Mg^{+2}(aq)	–614.55	–567.9	–286.8
MgO(s)	–998.57	–914.26	–282.76
MgH_2(s)	–658.3	–555.5	–346.99

(continued)

TABLE A.3 Standard-State Enthalpies, Free Energies, and Entropies of Atom Combination (continued)

Substance	ΔH°_{ac} (kJ/mol)	ΔG°_{ac} (kJ/mol)	ΔS°_{ac} (J/mol-K)
$Mg(OH)_2(s)$	–2005.88	–1816.64	–637.01
$MgCl_2(s)$	–1032.38	–916.25	–389.43
$MgCO_3(s)$	–2707.7	–2491.7	–724.2
$MgSO_4(s)$	–2708.1	–2448.9	–869.1
	Manganese		
Mn(s)	–280.7	–238.5	–141.69
Mn(g)	0	0	0
$Mn^{+2}(aq)$	–501.5	–466.6	–247.3
MnO(s)	–915.1	–833.1	–275.05
$MnO_2(s)$	–1299.1	–1167.1	–442.76
$Mn_2O_3(s)$	–2267.9	–2053.3	–720.1
$Mn_3O_4(s)$	–3226.6	–2925.6	–1009.7
$KMnO_4(s)$	–2203.8	–1963.6	–806.50
MnS(s)	–773.7	–695.2	–263.3
	Mercury		
Hg(l)	–61.317	–31.820	–98.94
Hg(g)	0	0	0
$Hg^{+2}(aq)$	+109.8	+132.58	–207.2
HgO(s)	–401.32	–322.090	–265.73
$HgCl_2(s)$	–529.0	–421.8	–359.4
$Hg_2Cl_2(s)$	–631.21	–485.745	–487.8
HgS(s)	–398.3	–320.67	–260.4
	Nitrogen		
$N_2(g)$	–945.408	–911.26	–114.99
N(g)	0	0	0
NO(g)	–631.62	–600.81	–103.592
$NO_2(g)$	–937.86	–867.78	–235.35
$N_2O(g)$	–1112.53	–1038.79	–247.80
$N_2O_3(g)$	–1609.20	–1466.99	–477.48
$N_2O_4(g)$	–1932.93	–1740.29	–646.53
$N_2O_5(g)$	–2179.91	–1954.8	–756.2
$NO_3^-(aq)$	–1425.2	–1259.56	–490.1
NOCl(g)	–791.84	–726.96	–217.86
$NO_2Cl(g)$	–1080.12	–970.4	–368.46
$HNO_2(aq)$	–1307.9	–1172.9	–454.5
$HNO_3(g)$	–1572.92	–1428.79	–484.80
$HNO_3(aq)$	–1645.22	–1465.32	–604.8
$NH_3(g)$	–1171.76	–1081.82	–304.99
$NH_3(aq)$	–1205.94	–1091.87	–386.1
$NH_4^+(aq)$	–1475.81	–1347.93	–498.8
$NH_4NO_3(s)$	–2929.08	–2603.31	–1097.53
$NH_4NO_3(aq)$	–2903.39	–2610.00	–988.8

TABLE A.3 Standard-State Enthalpies, Free Energies, and Entropies of Atom Combination (continued)

Substance	ΔH°_{ac} (kJ/mol)	ΔG°_{ac} (kJ/mol)	ΔS°_{ac} (J/mol-K)
$NH_4Cl(s)$	–1779.41	–1578.17	–682.7
$N_2H_4(l)$	–1765.38	–1574.91	–644.24
$N_2H_4(g)$	–1720.61	–1564.90	–526.98
$HN_3(g)$	–1341.7	–1242.0	–335.37
	Oxygen		
$O_2(g)$	–498.340	–463.462	–116.972
$O(g)$	0	0	0
$O_3(g)$	–604.8	–532.0	–244.24
	Phosphorus		
P(white)	–314.64	–278.25	–122.10
$P_4(g)$	–1199.65	–1088.6	–372.79
$P_2(g)$	–485.0	–452.8	–108.257
$P(g)$	0	0	0
$PH_3(g)$	–962.2	–874.6	–297.10
$P_4O_6(s)$	–4393.7	-	-
$P_4O_{10}(s)$	–6734.3	–6128.0	–2034.46
$PO_4^{3-}(aq)$	–2588.7	–2223.9	–1029
$PF_3(g)$	–1470.4	–1361.5	–366.22
$PF_5(g)$	–2305.4	-	-
$PCl_3(l)$	–999.4	–867.6	–441.7
$PCl_3(g)$	–966.7	–863.1	–347.01
$PCl_5(g)$	–1297.9	–1111.6	–624.60
$H_3PO_4(s)$	–3243.3	–2934.0	–1041.05
$H_3PO_4(aq)$	–3241.7	–2833.6	–1374
	Potassium		
$K(s)$	–89.24	–60.59	–96.16
$K(g)$	0	0	0
$K^+(aq)$	–341.62	–343.86	–57.8
$KOH(s)$	–980.82	–874.65	–357.2
$KCl(s)$	–647.67	–575.41	–242.94
$KNO_3(s)$	–1804.08	–1606.27	–663.75
$K_2Cr_2O_7(s)$	–4777.4	–4328.7	–1505.9
$KMnO_4(s)$	–2203.8	–1963.6	–806.50
	Silicon		
$Si(s)$	–455.6	–411.3	–149.14
$Si(g)$	0	0	0
$SiO_2(s)$	–1864.9	–1731.4	–448.24
$SiH_4(g)$	–1291.9	–1167.4	–422.20
$SiF_4(g)$	–2386.5	–2231.6	–520.50
$SiCl_4(l)$	–1629.3	–1453.9	–589
$SiCl_4(g)$	–1599.3	–1451.0	–498.03

(continued)

TABLE A.3 Standard-State Enthalpies, Free Energies, and Entropies of Atom Combination (continued)

Substance	ΔH°_{ac} (kJ/mol)	ΔG°_{ac} (kJ/mol)	ΔS°_{ac} (J/mol-K)
	Silver		
$Ag(s)$	–284.55	–245.65	–130.42
$Ag(g)$	0	0	0
$Ag^+(aq)$	–178.97	–168.54	–100.29
$Ag(NH_3)_2^+(aq)$	–2647.15	–2393.51	–922.6
$Ag_2O(s)$	–849.32	–734.23	–385.7
$AgCl(s)$	–533.30	–461.12	–242.0
$AgBr(s)$	–496.80	–424.95	–240.9
$AgI(s)$	–453.23	–382.34	–238.3
	Sodium		
$Na(s)$	–107.32	–76.761	–102.50
$Na(g)$	0	0	0
$Na^+(aq)$	–374.45	–338.666	–94.7
$NaH(s)$	–3811.25	–313.47	–228.409
$NaOH(s)$	–999.75	–891.233	–365.025
$NaOH(aq)$	–1044.25	–930.889	–381.4
$NaCl(s)$	–640.15	–566.579	–246.78
$NaCl(g)$	–405.65	–379.10	–89.10
$NaCl(aq)$	–636.27	–575.574	–203.4
$NaNO_3(s)$	–1795.38	–1594.58	–673.66
$Na_3PO_4(s)$	–3550.68	–3224.26	–1094.75
$Na_2SO_3(s)$	–2363.96	–2115.0	–803
$Na_2SO_4(s)$	–2877.20	–2588.86	–969.89
$Na_2CO_3(s)$	–2809.51	–2564.41	–813.70
$NaHCO_3(s)$	–2739.97	–2497.5	–808.0
$NaCH_3CO_2(s)$	–3400.78	–3099.66	–1013.2
$Na_2CrO_4(s)$	–2950.1	–2667.18	–949.53
$Na_2Cr_2O_7(s)$	–4730.6	-	-
	Sulfur		
$S_8(s)$	–2230.440	–1906.000	–1310.77
$S_8(g)$	–2128.14	–1856.37	–911.59
$S(g)$	0	0	0
$S^{2-}(aq)$	–245.7	–152.4	–182.4
$SO_2(g)$	–1073.975	–1001.906	–241.71
$SO_3(s)$	–1480.82	–1307.65	–580.3
$SO_3(l)$	–1467.36	–1307.19	–537.2
$SO_3(g)$	–1422.04	–1304.50	–394.23
$SO_4^{2-}(aq)$	–2184.76	–1909.70	–791.9
$SOCl_2(g)$	–983.8	–879.6	–349.50
$SO_2Cl_2(g)$	–1384.5	–1233.1	–508.39
$H_2S(g)$	–734.74	–678.30	–191.46

TABLE A.3 Standard-State Enthalpies, Free Energies, and Entropies of Atom Combination (continued)

Substance	ΔH°_{ac} (kJ/mol)	ΔG°_{ac} (kJ/mol)	ΔS°_{ac} (J/mol-K)
$H_2SO_3(aq)$	–2070.43	–1877.75	–648.2
$H_2SO_4(aq)$	–2620.06	–2316.20	–1021.4
$SF_4(g)$	–1369.66	–1217.2	–510.81
$SF_6(g)$	–1962	–1715.0	–828.53
$SCN^-(aq)$	–1391.75	–1272.43	–334.9
	Tin		
Sn(s)	–302.1	–267.3	–124.35
Sn(g)	0	0	0
SnO(s)	–837.1	–755.9	–273.0
$SnO_2(s)$	–1381.1	–1250.5	–438.3
$SnCl_2(s)$	–870.6	-	-
$SnCl_4(l)$	–1300.1	–249.8	–570.7
$SnCl_4(g)$	–1260.3	–1122.2	–463.5
	Titanium		
Ti(s)	–469.9	–425.1	–149.6
Ti(g)	0	0	0
TiO(s)	–1238.8	–1151.8	–306.5
$TiO_2(s)$	–1913.0	–1778.1	–452.0
$TiCl_4(l)$	–1760.8	–1585.0	–588.7
$TiCl_4(g)$	–1719.8	–1574.6	–486.2
	Tungsten		
W(s)	–849.4	–807.1	–141.31
W(g)	0	0	0
$WO_3(s)$	–2439.8	–2266.4	–581.22
	Zinc		
Zn(s)	–130.729	–95.145	–119.35
Zn(g)	0	0	0
$Zn^{2+}(aq)$	–284.62	–242.21	–273.1
ZnO(s)	–728.18	–645.18	–278.40
$ZnCl_2(s)$	–789.14	–675.90	–379.92
ZnS(s)	–615.51	–534.69	–271.1
$ZnSO_4(s)$	–2389.0	–2131.8	–862.5

Index